Recent Advances in Robotics, Factory Automation and Intelligent Networked Systems

Recent Advances in Robotics, Factory Automation and Intelligent Networked Systems

Guest Editor

Sergey Y. Yurish

MDPI

Basel • Beijing • Wuhan • Barcelona • Belgrade • Novi Sad • Cluj • Manchester

Guest Editor
Sergey Y. Yurish
International Frequency
Sensor Association (IFSA)
Castelldefels
Spain

Editorial Office
MDPI AG
Grosspeteranlage 5
4052 Basel, Switzerland

This is a reprint of the Special Issue, published open access by the journal *Machines* (ISSN 2075-1702), freely accessible at: https://www.mdpi.com/journal/machines/special_issues/industrialsystems_machines.

For citation purposes, cite each article independently as indicated on the article page online and as indicated below:

Lastname, A.A.; Lastname, B.B. Article Title. *Journal Name* **Year**, *Volume Number*, Page Range.

ISBN 978-3-7258-6364-8 (Hbk)
ISBN 978-3-7258-6365-5 (PDF)
https://doi.org/10.3390/books978-3-7258-6365-5

Contents

About the Editor

Sergey Y. Yurish

Sergey Y. Yurish has served as the president of the International Frequency Sensor Association (IFSA), one of the major professional associations serving the sensor industry and academia, since 1999. Dr. Yurish is a founder of three companies. He is Editor-in-Chief of the international peer-reviewed journal *Sensors & Transducers* and an editor for several open access multivolume Special Issue series. Dr. Yurish obtained his PhD degree in 1996 from the National University Lviv Polytechnic (UA). He has published more than 180 articles and papers in international peer-reviewed journals and conference proceedings. Dr. Yurish holds 9 patents and is the author and co-author of 12 books. He has delivered more than 90 speeches, tutorials and keynote presentations at industries, peer institutions, and professional conferences in 30 countries. Dr. Yurish was a Marie Curie Chair Excellence Investigator at the Technical University of Catalonia (UPC, Barcelona, Spain) from 2006 to 2009, where he led and developed one of the most successful projects in the UPC on Smart Sensors Systems Design (SMARTSES), totaling EUR 425,000. His professional accomplishments also include a Senior Research Fellowship at the Open University of Catalonia (UOC, Barcelona, Spain) where he spent a year in 2009-2010. Dr. Yurish has over 40 years of research and academic experience, during which he has developed numerous projects on an international level with various programmers, including NATO, FP6 and FP7. He is IARIA Fellow, and have achieved the Best Keynote Presentation Award at the 6th International Conference on Advances in Signal Processing and Artificial Intelligence (ASPAI' 2024).

Preface

This Reprint is devoted to recent advances in intelligent industrial systems, autonomous robotics, and data-driven automation. Its scope encompasses algorithms, architectures, and applications that enhance perception, decision making, manipulation, collaboration, and predictive capabilities in modern cyber–physical production environments. The aim of this collection is to consolidate emerging methodologies—from reinforcement learning and digital twinning to edge computing and privacy-preserving analytics—and to highlight their potential to address persistent industrial challenges related to safety, adaptability, efficiency, and interoperability.

The motivation for assembling this volume arises from the rapid evolution of Industry 4.0/5.0 technologies and the growing need for rigorous, experimentally validated research capable of guiding the next generation of smart manufacturing solutions. By bringing together contributions that span autonomous navigation, human–robot interaction, anomaly detection, energy modeling, and novel robotic applications, this Special Issue provides a coherent perspective on current trends and open problems in the field.

This reprint contains mainly the extended papers from the 5th IFSA Winter Conference on Automation, Robotics & Communications for Industry 4.0/5.0 (ARCI' 2025), organized by the International Frequency Sensor Association (IFSA) on the 19-21 February 2025, Granada, Spain. This work is addressed to researchers, engineers, practitioners, and graduate students engaged in robotics, automation, artificial intelligence, industrial engineering, and related disciplines. It is intended to serve both as a reference for current developments and as an inspiration for future research directions.

Sergey Y. Yurish
Guest Editor

machines

MDPI

Review

Robotic Applications in Skiing: A Systematic Review of Current Research and Challenges †

Răzvan Gabriel Boboc

Department of Automotive and Transport Engineering, Transilvania University of Brașov, 29 Eroilor, 500036 Brașov, Romania; razvan.boboc@unitbv.ro; Tel.: +40-745987710

† This paper is an extended version of our paper published in Proceedings of the 5th Winter IFSA Conference on Automation, Robotics & Communications for Industry 4.0/5.0 (ARCI' 2025), Granada, Spain, 19–21 February 2025.

Abstract: This paper provides a comprehensive review of the current state of research on robotic technologies in sports, with a specific emphasis on skiing. Using a systematic review methodology, I conducted an extensive search for relevant academic articles and conference papers across several databases, including Scopus, Web of Science, and IEEE. A predefined set of keywords guided the search process, leading to an initial collection of 327 papers. After applying specific selection criteria, 24 studies were identified as most relevant to the topic. These selected works were analyzed in detail, covering various key aspects of robotics in skiing, from biomechanical modeling to robotic ski systems. The findings underscore the growing interest in this interdisciplinary field while also highlighting significant research gaps, particularly in stability control, real-time terrain perception, and snow–surface interaction. Overall, the review concludes that while promising developments exist, the field is still in its early stages and would benefit greatly from interdisciplinary collaboration and more robust real-world experimentation.

Keywords: robotic technologies; skiing robotics; systematic review; humanoid robot; stability control

1. Introduction

Robots, as intelligent machines capable of sensing, processing information, and executing complex tasks, have transformed many industries and fields where human effort is required [1]. Advanced algorithms and control systems direct these sophisticated mechanical systems that represent the ultimate achievements of human engineering innovation. Although it is obvious that the world of sports and athletics has traditionally been dominated by the remarkable physical feats and capabilities of human beings, pushing the limits of what is possible through intense training, dedication, and willpower, a change has occurred in this sector as well. Specifically, in recent years, the emergence of advanced robotic systems has begun to redefine the limits of achievement in various sports fields, such as winter sports, especially skiing [2].

Given that skiing, a sport that requires exceptional balance, agility, and control, has historically been the exclusive domain of highly skilled human athletes, it is particularly interesting to note that researchers, inspired by this sport, have made efforts to develop robotic systems. These robots aim not only to imitate the movements and techniques of human skiers but also to surpass human performance in certain aspects [3].

One such example is the development of "skiing robots", which are autonomous systems that are capable of navigating complex ski slopes, carving turns, and even compet-

 https://doi.org/10.3390/machines13050397

ing in events like giant slaloms [4,5]. These robotic systems are equipped with advanced sensors, control systems, and algorithms that allow them to perceive their surroundings, maintain balance, and make rapid decisions while navigating the terrain. Thus, the potential of these robotic systems to improve skiing performance is high. They can not only perform the basic maneuvers of skiing but can also actively balance and control the pitch and roll of their skis to maintain stability on uneven surfaces [6]. This advanced control enables robots to navigate slopes more dynamically, mimicking the adaptive strategies used by human skiers.

Moreover, robot-based research offers several key advantages for studying skiing techniques. As highlighted in [7], the precise control over joint angles, forces, and movement trajectories allows researchers to program and fine-tune specific movement sequences, replicate the skiing styles and biomechanics of human athletes for in-depth performance analysis, develop and test new skiing techniques, and conduct controlled experiments on the influence of different ski equipment designs, materials, and settings. This approach not only accelerates the development of more efficient skiing methods but also supports advancements in sports science, equipment optimization, and even rehabilitation technologies for injured athletes looking to regain their skiing form.

Despite the growing interest in robotic applications in various sports fields, there are significant research gaps, particularly in the field of skiing robotics. While Erdmann has broadly outlined potential areas for the development of biomechanics-based sports robots [8], the specific application to ski technologies remains underdeveloped. In particular, there is limited research addressing the unique challenges of robotic locomotion on snow and ice surfaces, which present distinct friction and stability considerations compared to standard locomotion research.

Although Zheng et al. [9] have investigated humanoid robots walking on rough surfaces and proposed a ski-type gait, comprehensive studies specifically focused on robotic skiing mechanics are scarce. The dynamic nature of ski terrain requires specialized approaches to balance, trajectory planning, and energy efficiency that differ significantly from those used in traditional robotics applications. Furthermore, while Yin et al. [10] have demonstrated successful robotic applications in curling during the Beijing 2022 Winter Olympics, skiing presents significantly more complex biomechanical challenges due to its three-dimensional movement patterns, variable terrain, and higher speeds. The biomechanical modeling required for effective skiing robots has not been thoroughly explored in the literature.

Another notable gap is the lack of standardized testing methodologies for evaluating skiing robot performance. Unlike other robotic sports applications such as table tennis [11], where success metrics are well defined, skiing presents unique evaluation challenges due to its subjective performance elements and variable environmental conditions.

Finally, while technological advances continue to improve robotic capabilities, Riener [12] noted that humanoid robotics still lag behind human versatility and dexterity, particularly in complex locomotion tasks. This gap is especially pronounced in skiing, where the integration of perception, decision-making, and physical execution must occur rapidly in a dynamic environment.

Building on these identified gaps, this study aims to provide a perspective on research and innovative approaches for developing robotic solutions in skiing, addressing the unique challenges and the potential benefits of them for this winter sport. The paper is an extended version of "Enhancing Efficiency and Performance: Robotic Applications in Skiing" presented at the Fifth Winter IFSA Conference on Automation, Robotics, and Communications for Industry 4.0/5.0 (ARCI' 2025) [13]. The original work has been expanded with additional related work, research gap analysis, and a deeper exploration of robotic applications in skiing environments.

Despite growing interest in robotic applications in sports, the specific field of skiing robotics remains fragmented and lacks a consolidated overview of its technical challenges, research gaps, and practical potential. The motivation behind this review is to systematically synthesize the available knowledge, highlight the critical issues limiting progress, and outline concrete application areas and development directions. In this way, this study aims to provide a structured foundation for future research efforts, bridging gaps between robotics, biomechanics, sports engineering, and rehabilitation technologies related to skiing.

The structure of this paper is as follows: Section 2 reviews prior work related to robotic technologies in sports. Section 3 outlines the systematic methodology used to identify and select relevant studies. Section 4 presents and discusses the results, including publication trends, research topics, robotic systems, and identified challenges. Finally, Section 5 concludes this paper by summarizing key findings and outlining future directions for research in skiing robotics.

2. Related Work

2.1. Robotics in Sports: An Overview

The intersection of robotics and sports has gained considerable attention in recent years. Erdmann outlined various applications of robots in sports, including serving balls, assisting in training, substituting humans during practice, participating in competitions, serving as models for performance analysis, and supporting event organizers [8]. This work underscores the transformative potential of robotics across multiple facets of sports.

Building on this, Siegel and Morris explored how robotics and automation are redefining the sports experience—not only enhancing athlete performance but also revolutionizing fan engagement and paving the way for entirely new competitive formats [14]. They examined diverse applications, including immersive fan experiences, automated refereeing, robotic athletes, and advancements in motorsports. Their analysis balanced technological innovation with the importance of preserving the emotional, human essence that makes sports captivating.

Hu et al. contributed to this growing field by focusing on game motion robots powered by gait recognition algorithms [15]. Their research demonstrated how such systems can capture detailed, real-time data on athlete movements, improving training precision and playing a key role in injury prevention. This approach reflects a shift toward using robotics not merely as training aids but as sophisticated, data-driven performance enhancers.

Together, these studies reveal a future where robotics in sports extends beyond automation, evolving into an intelligent, adaptive force that supports athletes, enriches fan experiences, and keeps the spirit of competition alive.

2.2. Robotic Locomotion on Ice and Snow

Research on robotic locomotion on slippery surfaces provides valuable insights for skiing robotics. Iverach-Brereton et al. described the development of a gait to allow a small humanoid robot to propel itself on ice skates across a smooth surface [16]. Their work with both ice skates and inline skates demonstrated that a new gait relying entirely on motion in the frontal plane allowed the robot to traverse ice surfaces more stable than classic inverted pendulum-based walking gaits when using the same skates.

Yin et al. designed and manufactured a hexapod curling robot that imitates human kicking, sliding, pushing, and curling rock rotating, which completed a successful demonstration during the Beijing 2022 Winter Olympics Games [10]. This robot's morphology transformation based on leg configuration switching allowed smooth transitions between walking and throwing tasks, demonstrating potential applications for adaptive robotic systems in winter sports environments.

Zheng et al. [9] investigated how humanoid robots can walk on rough surfaces, proposing new walking patterns including a "step-over" gait and a ski-type gait as global approaches to maintain stability while using compliant motion to solve the robust foot-holding problem. This research provides foundational approaches that could be adapted specifically for skiing robots.

2.3. Philosophical and Comparative Perspectives

The potential participation of robots in sports raises interesting philosophical questions. In this way, Lopez Frias and Pérez Triviño addressed the philosophical and ethical implications, drawing on existing discussions about the nature of sports and athletic skill [17]. The authors argue that for robots to truly engage in sports, they would need (a) the ability to comprehend their environment and exhibit intentional states; (b) a drive to excel and a playful, competitive spirit, a "ludic attitude" that embraces the joy of overcoming challenges; and (c) a physical form that embodies vulnerability, making them admirable, much like human athletes. Furthermore, ethical concerns about fairness arise when considering the direct competition between robots and human athletes. Robots, potentially exceeding human physical capabilities through optimized design and artificial intelligence, could challenge the fundamental principles of equal opportunity and natural human performance that underpin competitive sports [17,18].

Building on this, the notion of robots truly comprehending their environment and exhibiting intentional states is complex. While current entertainment fitness robots can perceive a user's actions and respond accordingly using action recognition [19], and robots in development for sports like soccer utilize sensors to perceive their environment, the question of genuine intentionality remains. Riener et al. note that robots can move, perceive, and communicate but currently lack an emotional spectrum, such as feelings of happiness or frustration [12], which are deeply intertwined with human intentionality and comprehension. Trenchard argues that for robots to truly demonstrate human-level intelligence, they must exhibit self-organized behaviors like the invention of team sports, which implies a level of understanding beyond mere programmed responses [18]. This would necessitate robots deciding to create their own games involving competition and cooperation, potentially arising from an intrinsic capacity to compete and cooperate for resources and sufficient leisure time.

In terms of physical design, while robotic actuators can surpass human muscles in specific functions, they often do so at the expense of added size and weight, which can reduce agility and versatility [12]. Additionally, human athletic ability is shaped by unique biological and evolutionary processes, influencing how we perceive and admire physical skill. For robots to attain human-level intelligence, they must evolve through similar conditions and processes as humans. Their physical existence and interactions are essential in shaping the philosophical understanding of their "athleticism" [18].

This growing body of interdisciplinary research—spanning robotics, biomechanics, machine learning, and sports science—sets the foundation for a structured exploration of how robotics can be applied specifically to skiing. The following section outlines the systematic methodology used to identify, evaluate, and synthesize relevant studies in this domain.

2.4. Robotic Technologies in Skiing

The use of robotics in skiing is a niche yet rapidly advancing research domain. Early explorations focused on passive robotic systems that mimicked basic skiing maneuvers for understanding the dynamics of carved turns [7]. More recent advancements involve active skiing robots equipped with sophisticated control algorithms, perception systems, and articulated joints to navigate complex ski slopes and replicate or even enhance human skiing techniques.

Notable developments include humanoid robots like RoK-2 and DIANA, which are designed to autonomously ski through slalom gates, using real-time vision systems and stability control frameworks to maintain dynamic balance [4,20]. These robots typically integrate LiDAR sensors, vision modules, and inertial measurement units (IMUs) to perceive terrain and adjust body posture accordingly.

Researchers have proposed various control strategies to ensure stability during high-speed descents. Methods such as Zero Moment Point (ZMP) control and hybrid reflexive mechanisms are used to stabilize humanoid robots in unpredictable terrain conditions [5,21]. The use of reinforcement learning for adaptive control has also emerged, enabling robots to learn optimal skiing strategies through trial-and-error in simulated environments before real-world deployment [22].

A major research focus is the replication of human skiing biomechanics in robotic systems. Movements such as knee flexion–extension and hip abduction–adduction are critical for balance and maneuverability during skiing. Robotic platforms attempt to reproduce these motions using actuated joints and compliant mechanisms [4,7,23].

In alpine skiing, robots must adapt to changes in slope gradient and snow texture. Lower-body robotic platforms have been developed to fine-tune ski edge angles and weight distribution, key for executing carved turns. These robots use servo motors to emulate joint kinematics and maintain posture during dynamic turns, closely mirroring human biomechanics.

Autonomous skiing robots must detect and respond to obstacles such as gates, terrain irregularities, and other skiers. Vision-based systems, often enhanced by deep learning algorithms, are employed for real-time gate recognition and path planning [20]. Multi-modal sensor fusion, combining LiDAR and computer vision, improves environmental awareness and enables precise trajectory adjustments during descent.

Obstacle avoidance strategies have been proposed for various skiing scenarios, including freestyle and alpine racing. For example, six-legged skiing robots demonstrate enhanced stability and versatility across flat, downhill, and uphill terrains, using gait adaptation to maneuver around dynamic obstacles [24].

2.5. Novel Robotic Platforms

Recent review studies have highlighted key trends and challenges in the development of mobile and sports-oriented robotic systems. García and Duarte analyzed the adaptation of mobile ground robots in outdoor environments, identifying terrain variability and real-time perception as critical challenges [25]. Hu et al. surveyed the evolution of game sports robots, emphasizing performance optimization in dynamic conditions [15]. Ju et al. discussed broader trends in sports rehabilitation robotics for elderly patients [26]. Wu et al. presented the theoretical research and engineering development of legged robots by examining key technologies, walking mechanisms, dominant models, and techniques and discussing applications and challenges [27]. Another review paper explored the design, locomotion, actuation, and control of soft-legged robots for diverse environments, highlighting their adaptability, efficiency, and potential in rescue, military, and space applications [28].

Building upon these insights, several novel robotic platforms have emerged. Bakırcıoğlu et al. [29] and Chambers et al. [30] developed hydraulic and adaptive leg mechanisms that enhance mechanical robustness and energy efficiency, essential for stability on slopes.

Chen et al. [31] and Cui et al. [32] proposed soft and tensegrity-based robots, respectively, offering compliance and resilience suitable for variable snow conditions. Marcus et al. [33] presented a novel walking algorithm for quadruped robots that use reinforcement learning for adaptable movement in hazardous, unstructured environments. Wang et al. [34] developed a six-legged skiing robot that emulates human techniques for stable cross-country skiing on complex snow, using terrain estimation and balance control.

AI and control innovations have also advanced robotic capabilities. Afzal et al. explored the potential of integrating artificial intelligence with animal-inspired robots to revolutionize wildlife study and enable more effective ecological conservation strategies [35]. Nguyen and Kim [36] introduced fixed-time control methods for rapid stability achievement. Xue et al. [14] developed a force-position hybrid control for quadruped robots to enhance stability and adaptability on complex terrain [37]. Khan et al. presented a multi-agent reinforcement learning controller for a wheel-legged robot to enable autonomous learning of various gaits and robust control [38].

Exoskeletal and modular robotic designs have been introduced to enhance adaptability. Jamali et al. developed a portable hybrid soft-cable knee exoskeleton for rehabilitation [39], while Ma et al. presented a quantitative method for evaluating lower limb exoskeleton training effectiveness by integrating electrophysiological and kinematic data through synergistic indicators [40]. Marchesi et al. contributed to rehabilitation robotics, emphasizing trunk control, which could inform the development of balance-assisting skiing robots [41].

Li et al. proposed a dynamic modeling strategy to enhance the performance of industrial robots [42], while Zhang et al. introduced a Bayesian kinematic calibration method to identify joint variables in industrial robots while accounting for uncertainty, improving positioning accuracy and enabling robust trajectory predictions [43].

This growing body of interdisciplinary research, spanning robotics, biomechanics, machine learning, and sports science, sets the foundation for a structured exploration of how robotics can be applied specifically to skiing. The following section outlines the systematic methodology used to identify, evaluate, and synthesize relevant studies in this domain.

3. Methodology

The systematic review was designed to ensure a comprehensive and transparent exploration of the role of robotics in skiing. Following the PRISMA guidelines [44], the methodology ensured clarity and reproducibility, covering all stages, from the initial search to data extraction and analysis, to capture the most relevant and high-impact studies on this emerging topic.

3.1. Search Strategy

This literature search was conducted on 26 August 2024, utilizing three leading electronic databases: Scopus, Web of Science, and IEEE Xplore, chosen for their extensive coverage of scientific and technical literature. To cast a wide net while ensuring precision, this search employed a combination of keywords and Boolean operators:

(robot* OR robotics) AND (skiing OR ski OR "ski training")

These terms targeted the title, abstract, and keywords fields, ensuring the inclusion of articles explicitly mentioning robotics within the context of skiing. The wildcard symbol (*) captured variations of "robot"—including "robots" and "robotics"—to accommodate diverse terminology across disciplines and regions.

This approach aimed to capture a broad spectrum of studies, ranging from theoretical explorations and algorithmic advancements to practical implementations of robotic systems in skiing. The intent was not only to gather data on existing robotic technologies but also to identify research gaps and future directions.

3.2. Eligibility Criteria

To ensure the inclusion of high-quality and contextually relevant studies, strict eligibility criteria were established:

- Inclusion criteria:
    - Studies focusing on robots or robotic systems specifically within skiing contexts (e.g., training, performance analysis, or autonomous skiing);
    - Peer-reviewed journal articles or conference proceedings, ensuring scientific rigor and credibility;
    - No restrictions on publication date, allowing the review to encompass both foundational studies and the latest advancements.
- Exclusion criteria:
    - Non-research documents, such as books, reviews, editorials, or communications, that lacked original empirical data;
    - Duplicate records retrieved from multiple databases;
    - Papers unrelated to skiing or studies that mentioned robotics but lacked direct application to skiing;
    - Non-English articles or studies unavailable in full text, ensuring accessibility and comprehensibility.

This balanced approach preserved the integrity of the review by prioritizing scientific contributions while filtering out irrelevant, redundant, or inaccessible content.

3.3. Study Selection Process

The selection process unfolded in a series of systematic stages designed to progressively refine the dataset:

1. Initial search results: A total of 327 documents matched the search criteria across all databases.
2. Duplicate removal: 67 duplicate entries were identified and excluded to prevent redundancy, ensuring each study was counted only once.
3. Title and abstract screening: Next, a preliminary screening of titles and abstracts was performed. Articles that failed to mention robotics or skiing in a meaningful, research-driven context were eliminated. This step resulted in the exclusion of 198 papers, many of which referenced skiing metaphorically or focused on unrelated robotic applications.
4. Full-text review: The remaining 62 articles underwent an in-depth, full-text evaluation. This stage assessed the relevance, scientific rigor, and accessibility of each study. A total of 38 papers were excluded due to the following reasons:
    - Emphasis on unrelated robotic applications or sports;
    - Full-text inaccessibility;
    - Non-English language;
    - Being classified as review papers rather than original research.

At the conclusion of this process, 24 articles met the eligibility criteria, offering a robust and focused body of literature to support this study's objectives.

Figure 1 visually presents the PRISMA-based study selection process, illustrating the systematic progression from initial search to final inclusion.

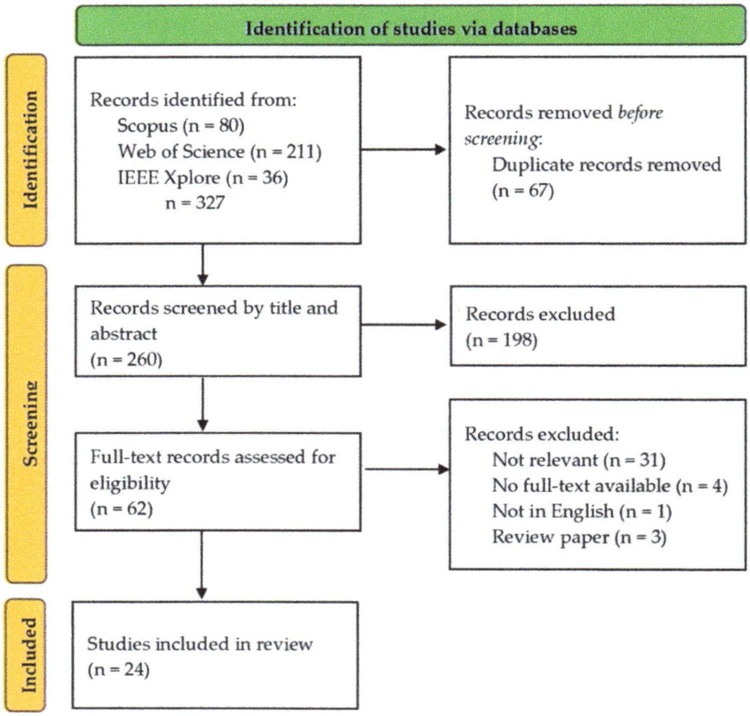

Figure 1. PRISMA-based research process diagram.

3.4. Data Extraction and Analysis

The final stage involved extracting detailed data from the remaining 24 articles. To ensure accuracy and consistency, EndNote 20.6 was employed for reference management and citation tracking. Each article was scrutinized, and key information was transferred into a structured Excel spreadsheet for streamlined analysis.

The extracted data encompassed years of publication, sources, authors, affiliations, citations, keywords, study objectives and topics, robotic system types, technologies, software, and limitations.

This structured approach allowed for comprehensive cross-study comparisons, highlighting recurring themes, emerging trends, and unresolved challenges.

3.5. Quality and Bias Assessment

Given the niche nature of robotics in skiing, the current body of literature remains limited. As a result, this study prioritized capturing a comprehensive overview of existing research rather than applying strict quality and bias assessments, which could have further reduced the already small pool of eligible studies. Implementing standardized evaluation frameworks, such as the Critical Appraisal Skills Program (CASP) [45] or Cochrane Risk of Bias (RoB) [46], might enhance the reliability of future reviews. However, in this context, such criteria would likely have excluded a significant portion of the remaining studies, potentially limiting insights into this emerging field.

While this review refrains from formal quality scoring, the inclusion criteria ensured that only peer-reviewed, research-based contributions were selected, preserving a baseline level of academic rigor. Future research efforts should aim to expand the knowledge base, allowing for more comprehensive quality assessments without the risk of losing valuable, albeit preliminary, findings.

4. Results and Discussion

A comprehensive database was developed based on the results that met the established inclusion criteria. This database was structured to capture key information across several categories: (i) the year of publication, (ii) the type of publication (journal or conference), (iii) the country of origin of the research, (iv) the authors and their respective affiliations, (v) citation counts to assess the impact of the papers, (vi) the keywords used in the studies, (vii) the main research topics covered, (viii) the specific aims of the research, and (ix) the types of robots involved in the studies. This systematic approach allowed for a detailed and organized overview of the relevant literature in the field.

4.1. Years of Publication and Sources

Figure 2 offers a chronological visualization of the reviewed papers, categorizing them by year of publication and type (journal or conference proceeding). The timeline begins with the earliest identified work from 2009, which also emerged as the most prolific year, with five publications—a notable surge in comparison to other years. Following this, the field saw a steady trickle of one to two publications annually, reaching a minor peak in 2020 with three papers. It is worth highlighting that this analysis reflects data gathered by August 2024; any research published after this date was not included.

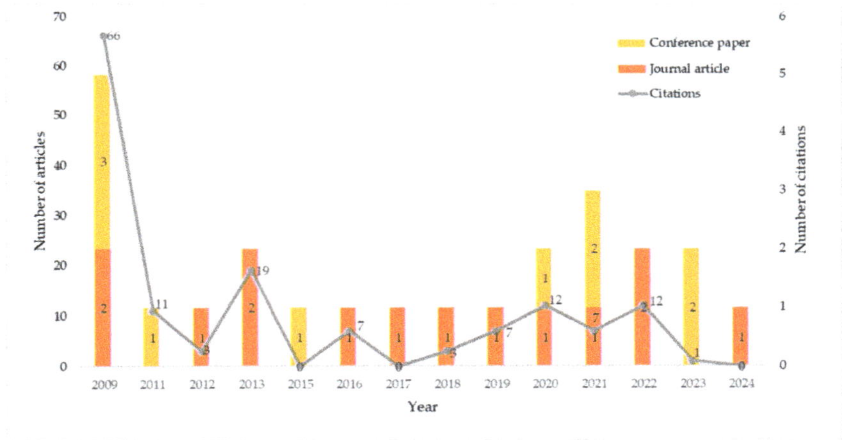

Figure 2. Distribution of analyzed papers by publication type and year.

Additionally, Figure 2 illustrates the total number of citations these papers accumulated per year, providing insight into the broader academic influence and visibility of the topic over time.

In terms of publication sources, the majority of contributions (14 articles, representing 58.33%) appeared in peer-reviewed journals, reflecting a baseline of academic rigor. The remaining 10 studies (41.67%) were published in conference proceedings, indicating active discussion within scientific communities. Among journals, *Sensors* emerged as the most frequent publisher, contributing three articles to the field. On the conference side, the IEEE/RSJ International Conference on Intelligent Robots and Systems and the International Conference on Intelligent Robotics and Applications each featured two papers across different years, while other conferences contributed one article each.

This distribution highlights that, despite the relatively small number of studies, research on skiing robotics spans reputable and diverse publication venues, combining theoretical advances with practical developments discussed at leading robotics conferences. Table 1 presents an overview of the selected articles based on their publication year, country of origin, and source names.

Table 1. Summary of selected articles by publication year, country, and source.

No.	Ref.	PY	Country	Source
1	[4]	2020	Republic of Korea	*International Journal of Precision Engineering and Manufacturing*
2	[47]	2021	Republic of Korea	ICECCIT 2021 Proceedings
3	[48]	2020	France	CCTA 2020 Proceedings
4	[6]	2016	Canada	*The Knowledge Engineering Review*
5	[5]	2019	Republic of Korea	*Sensors*
6	[23]	2012	Japan	*Journal of Robotics and Mechatronics*
7	[49]	2009	Slovenia	*Robotica*
8	[3]	2009	Slovenia	IEEE/RSJ International Conference on Intelligent Robots and Systems
9	[50]	2021	Austria	*Frontiers in Sports and Active Living*
10	[20]	2022	Republic of Korea	*Sensors*
11	[21]	2011	Slovenia	*Autonomous Robots*
12	[51]	2013	Slovenia	IEEE/RSJ International Conference on Intelligent Robots and Systems
13	[52]	2013	Slovenia	*International Journal of Humanoid Robotics*
14	[53]	2009	Italy	ASME International Mechanical Engineering Congress and Exposition
15	[54]	2009	Japan	*Applied Sciences*
16	[55]	2018	Japan	*Lecture Notes in Computer Science*
17	[56]	2022	Austria	IEEE Sensors 2023
18	[57]	2023	Austria	*Sensors*
19	[58]	2017	China	*Acta Technica*
20	[22]	2021	China	International Conference on Security, Pattern Analysis, and Cybernetics
21	[24]	2023	China	*Lecture Notes in Computer Science*
22	[59]	2024	China	*IEEE Robotics and Automation Letters*
23	[7]	2009	Japan	*Sports Engineering*
24	[60]	2015	China	IEEE International Conference on Industrial Technology Proceedings

Note: PY—publication year.

4.2. Countries, Authors, and Institutions

The documents were further categorized based on their country of origin. Figure 3 provides a visual representation of the geographical distribution of the 24 studies included in the systematic review, considering the countries of all authors involved in each study. According to the analysis, Slovenia and China stand out as the leading contributors to research in this area, each accounting for five publications. South Korea and Japan follow closely with four publications each, while Austria contributes three publications. This geographical breakdown highlights the countries most actively engaged in advancing research on robotics in skiing.

An important aspect of the analysis focused on identifying the most prolific authors, based on the number of publications they contributed to in the field. Figure 4 provides a visual representation of the author's distribution, highlighting those who have significantly contributed to the research on robotics in skiing. Notably, Bojan Nemec emerged as the author with the highest number of publications, with a total of four studies. Close behind, Tadej Petrič, Norihiko Saga, and Leon Žlajpah each contributed three publications. When considering first authorship specifically, Tadej Petrič stood out as the most frequent first author, with three studies attributed to him. This analysis offers valuable insights into the key contributors driving research in this emerging area.

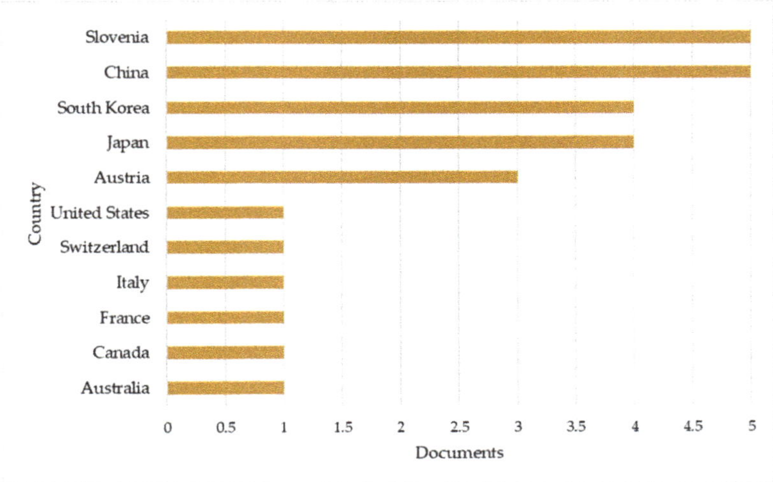

Figure 3. Geographical distribution of papers based on authors' countries of origin.

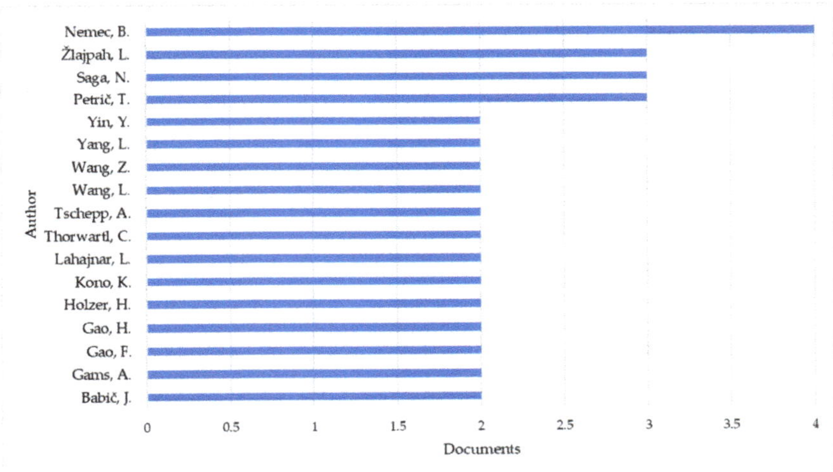

Figure 4. Distribution of publications by authors with two or more contributions.

Figure 5 presents the distribution of institutions affiliated with the authors of the reviewed papers. The Jožef Stefan Institute stands out as the leading institution, contributing a total of five publications to the field. Following closely is Kwansei Gakuin University, with three published studies. Several other institutions have made notable contributions as well, with Atomic Austria GmbH, Red Bull Athlete Performance Center, Universität Salzburg, Shanghai Jiao Tong University, and Joanneum Research Forschungsgesellschaft mbH each being affiliated with two publications. This distribution highlights the prominent role of these institutions in advancing research in the area of robotics applied to skiing.

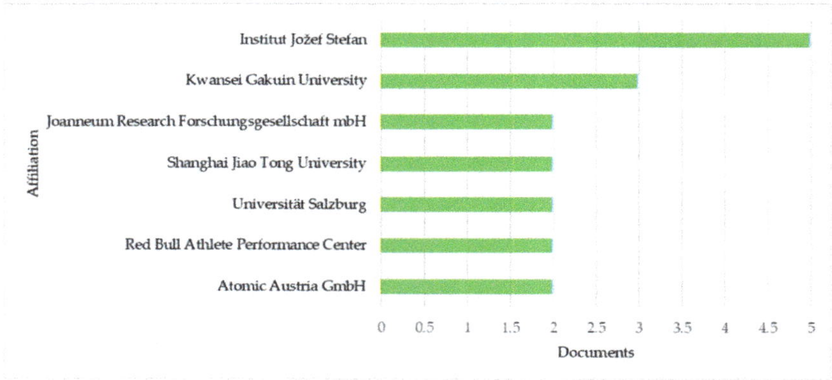

Figure 5. Institutional affiliations of the published papers.

4.3. Citations, Keywords, and Research Aims

To assess the impact of the research in this field, the number of citations for each article was analyzed. Table 2 presents the top 10 most cited articles in descending order, with citation counts sourced from the Scopus database. The article with the highest number of citations, at 23, is "Design, control, and navigation in unstructured environments", published in *Robotica* in 2009. This is followed by "Control and navigation of the skiing robot", presented at the 2009 IEEE/RSJ International Conference on Intelligent Robots and Systems, and "A ski robot system for qualitative modelling of the carved turn", published in *Sports Engineering* in 2009. These articles represent the most influential contributions to the development of robotic systems in skiing research.

Table 2. Top 10 most cited works.

No.	Ref.	Source	No. of Citations
1.	[49]	*Robotica*	23
2.	[3]	IEEE/RSJ IROS 2009	21
3.	[7]	*Sports Engineering*	19
4.	[21]	*Autonomous Robots*	16
5.	[51]	IEEE/RSJ IROS 2011	11
6.	[4]	*International Journal of Precision Engineering and Manufacturing*	7
7.	[6]	*Knowledge Engineering Review*	7
8.	[5]	*Sensors*	6
9.	[57]	*Sensors*	5
10.	[20]	*Sensors*	4

Figure 6 presents a visual representation of keyword co-occurrences, generated using VOSviewer version 1.6.20. A minimum occurrence threshold of two was set, and from a total of 232 keywords, 38 met this criterion. The resulting map is divided into five distinct clusters, with a total of 162 links between the keywords.

Cluster 1 (red): this cluster includes terms such as "robots", "degrees of freedom", "sensory systems", and "machine design", which highlight the technological and mechanical aspects of robotics.

Cluster 2 (yellow): keywords in this cluster focus on "humanoid robot", "deep learning", and "human-robot interaction", indicating a strong connection between artificial intelligence, robotics, and user interaction.

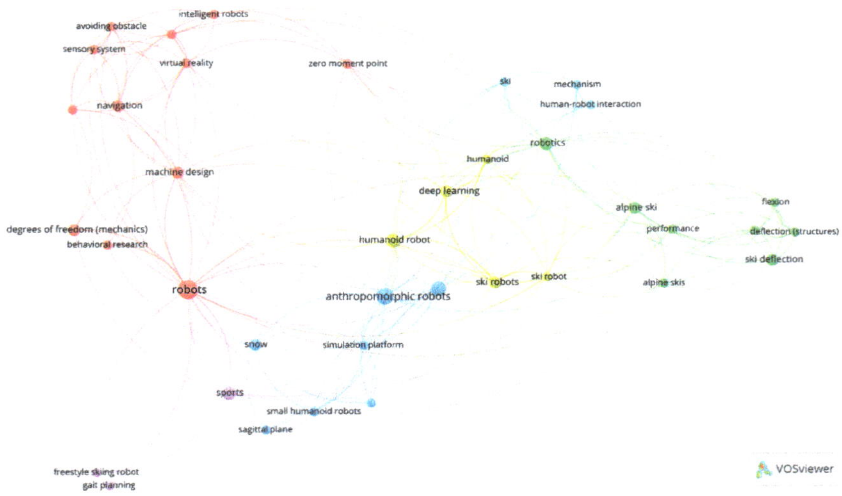

Figure 6. VOSViewer network map of co-occurring keywords.

Cluster 3 (green): this group revolves around "alpine ski", "ski deflection", and "performance", emphasizing the application of robotics in the context of skiing and related performance metrics.

Cluster 4 (blue): keywords like "anthropomorphic robots", "stability control", and "skiing robot" reflect research on robots designed to simulate human-like movements and maintain balance during skiing.

Cluster 5 (purple): this cluster includes "sports", "freestyle skiing robot", and "gait planning", linking robotics to the broader sports domain, with a focus on freestyle skiing and the planning of movement patterns.

Notable keyword occurrences include: "robots" (eight occurrences, total link strength 32), "anthropomorphic robots" (six occurrences, total link strength 32), "skiing robot" (five occurrences, total link strength 17), "humanoid robot" (four occurrences, total link strength 24), "robotics" (four occurrences, total link strength 20), and "sports" (four occurrences, total link strength 16).

This analysis underscores the key areas of focus within the field, particularly the intersection of robotics and skiing, as well as the growing emphasis on human–robot interaction and artificial intelligence in robotic design.

Tables 3 and 4 provide a comprehensive summary of the search results, detailing key aspects of each selected study. It outlines the primary research objective of each paper, alongside a description of the robots employed, including their specific type and configuration. Additionally, the table highlights the degrees of freedom (DOF) associated with each robotic system, offering insight into their range of motion and adaptability. It also categorizes the type of experiment conducted (e.g., simulation, laboratory setup, or real-world trials), providing a clearer understanding of how robots were tested and validated within the context of the research.

To better reflect the significant differences in application purposes, the studies have been divided into two categories: (i) skiing robots aimed at dynamic movement on slopes, and (ii) industrial robots aimed at testing ski-related equipment parameters. For both categories, where such information was found, additional technical specifications regarding sensors, control systems, and software environments have been included to provide a more detailed technological overview.

Table 3. Review of selected papers (skiing robots).

No.	Ref.	Research Aim	Robot/Type	Experiment Type	DOFs	Sensors	Control Subsystems	Software Tools
1	[4]	Develop a humanoid robot capable of autonomously performing in a ski giant slalom competition.	RoK-2 humanoid robot/SR	real	20	encoders, IMU, 3-axis FT sensors, video camera	image processing unit, deep learning-based object detection, stability control	N/A
2	[47]	Propose a trajectory planning algorithm for a multi-degree-of-freedom robot using CSP mode.	TiBo/SR	simulation (virtual)	NS	N/A	N/A	N/A
3	[48]	Develop a high-degree-of-freedom simulator for sliding sports.	Evr@ simulator/RP	real	6	displacement, velocity, and torque sensors of the PD4 motors, infrared camera	trajectory planning, virtual obstacle avoidance	Matlab
4	[6]	Develop a simple control system for an alpine skiing robot.	DARwIn-OP/SR	real	12	N/A	N/A	N/A
5	[5]	Propose a general stability control method for a bipedal alpine skiing robot using ZMP and LiDAR.	DARwIn-OP/SR	simulation	12	LiDAR	stability control, turning algorithm	Webots, Matlab
6	[23]	Develop a passive skiing robot to investigate the mechanisms of ski turns.	Passive turn skiing robot/SR	real	1	N/A	N/A	N/A
7	[55]	Develop a passive skiing robot to investigate the mechanisms of ski turns.	Passive turn skiing robot/SR	real	3	speed sensor, magnetic field sensor, wheel encoders, video camera	carved turn control	N/A
8	[54]	Develop a passive skiing robot to facilitate understanding of the turn mechanism for junior skiers.	Passive turn skiing robot/SR	real	3	small servomotor	turn control	N/A
9	[49]	Develop a skiing robot capable of autonomous navigation on a ski slope.	Planar leg robot/SR	real + virtual	6	electronic gyroscope, force sensors, motor position sensors, GPS receiver, camera	stability control, navigation, decision making	virtual reality environment
10	[3]	Develop a skiing robot capable of autonomous navigation on a ski slope using carving technique.	Planar leg robot/SR	real	23	electronic gyroscope, force sensors, motor position sensors, GPS receiver, camera	stability control, navigation, decision making	N/A

Table 3. *Cont.*

No.	Ref.	Research Aim	Robot/Type	Experiment Type	DOFs	Sensors	Control Subsystems	Software Tools
11	[21]	Propose a general control framework for ensuring stability of humanoid robots using ZMP.	Planar leg robot/SR	real	3 (+6)	motor position sensors, force sensors, electronic gyroscope	stability control	N/A
12	[51]	Propose a novel method for ensuring stability of a skiing robot on an unknown ski slope.	Planar leg robot/SR	real	3 (+6)	motor position sensors, force sensors, electronic gyroscope	stability control	N/A
13	[52]	Propose and evaluate methods for local navigation of a skiing robot using visual perception.	Planar leg robot/SR	real	3	motor position sensors, force sensors, electronic gyroscope	stability control, navigation	N/A
14	[20]	Develop a humanoid robot capable of autonomously performing in a ski giant slalom competition.	DIANA/SR	real	2	stereo camera, LiDAR sensors, F/T sensors, IMU sensors	path planning, vision recognition, carved turn control, stability control	ROS, Matlab
15	[58]	Develop a freestyle skiing robot model and gait planning based on human body data.	Virtual prototype/SR	simulation	NS	N/A	gait planning	ADAMS
16	[22]	Develop a skiing robot capable of independently performing skiing movements using DDPG reinforcement learning.	Humanoid robot/SR	real	12	N/A	N/A	N/A
17	[59]	Develop an obstacle avoidance method for skiing robots.	Six-legged skiing robot/SR	real	5	IMU, LiDAR	locomotion controller, turn control	N/A
18	[24]	Design, control, and apply a novel six-legged skiing robot on flat and slope terrains.	Six-legged skiing robot/SR	real	5	IMU, LiDAR	trajectory planning, vbalance control, turn control	NX
19	[7]	Develop a robot that simulates human leg joint motions during carved turns.	Robotic system/SR	real	6×2	strain gauges, wheel encoders, pulse count sensor, magnetic field sensor	turn control	N/A
20	[60]	Develop a freestyle skiing robot model and gait planning based on human body data.	Virtual prototype/SR	simulation	24	gyroscope, accelerometer, six-axis force sensor	stability control, gait controller	ADAMS

Note: SR—skiing robot, RP—robotic platform, DOFs—degrees of freedom, IMU—inertial measurement unit.

Table 4. Review of selected papers (industrial testing robots).

No.	Ref.	Research Aim	Robot/Type	Experiment Type	DOFs	Sensors	Control Subsystems	Software Tools
1	[50]	Develop a new method to release ski bindings using an industrial robot.	KUKA Quantec series/RA	real	1	PyzoFlex® sensors, optoelectrical measuring system	motion pattern recognition, release force determination	Matlab
2	[53]	Propose a novel device and method to evaluate the mechanical behavior of ski boot stiffness.	XY Cartesian robot/CR	real	1	force and position sensors	ski boot stiffness calculation	N/A
3	[57]	Test the performance of an enhanced ski deflection measurement prototype in dynamic settings.	IRB 6400R, ABB AG/RA	real	6	PyzoFlex® sensors, 12 sensor foils, optoelectrical measuring system, eight active infrared cameras	measurement data acquisition, synchronization unit	Matlab
4	[56]	Evaluate the feasibility of detecting torsional deflections in alpine skis using a parallel sensor layout.	IRB 6400R, ABB AG/RA	real	6	PyzoFlex® sensors, 12 sensor foils, optoelectrical measuring system, eight active infrared cameras	measurement data acquisition, synchronization unit	Matlab

Note: RA—robotic arm, CR—cartesian robot, DOFs—degrees of freedom.

4.4. Key Technical Domains

The analysis also considered the primary research topics addressed in the selected studies. Figure 7 provides an overview of these topics, revealing that the most frequently explored subject is carved turn control and stability control followed by obstacle avoidance, autonomous navigation, and detection of ski deflection.

Figure 7. Sankey diagram illustrating the relationship between robot types and the research topics explored in the studies.

4.4.1. Robotic Morphologies and Hardware Design

Regarding the robotic platforms utilized, the majority of studies (79%) focused on skiing robots, with lower-body platforms and humanoid robots being the most represented category, appearing in ten studies. The classification of robot types includes the following: lower body mechanical platforms (six studies), humanoid robots (four studies), virtual prototypes (four studies), passive turn mechanisms (three studies), industrial articulated-arm robots (three studies), six-legged robots (two studies), simulation platforms (one study), and cartesian robots (one study). It is important to note that a single study may cover multiple topics.

Overall, the reviewed studies highlight the diversity of robotic systems employed in skiing research, ranging from humanoid robots and simulation platforms to specialized robotic mechanisms such as six-legged skiing robots and industrial robots used for evaluating ski equipment performance.

4.4.2. Environmental Perception

Various techniques have been employed for object recognition in skiing environments, particularly for detecting flags, gates, or other obstacles on the slope. These methods range from traditional image processing techniques [3,49] to more advanced artificial intelligence-driven approaches. Deep learning algorithms have been leveraged to enhance recognition accuracy [4], while ranging sensors such as LiDAR have been utilized for precise distance measurement and obstacle detection [5,24,59]. Additionally, visual perception systems integrated with neural networks have been implemented to improve real-time recognition and response capabilities [52]. A more sophisticated approach combines convolutional neural networks (CNNs) with LiDAR sensors, providing a robust multi-modal perception framework for enhanced environmental awareness and object identification on the ski slope [20].

4.4.3. Balance Control and Stability Mechanisms

Research has also demonstrated that humanoid robots can effectively replicate carved turns through coordinated leg movements, including extension–flexion and abduction–adduction [4,7].

In the sagittal plane, extension–flexion has a significant impact on skiing dynamics. When a skier crouches, flexing the knees and hips, they absorb shocks and lower their center of mass for stability. Conversely, extending the legs propels the skier into the next turn, generating power and control. For humanoid robots, servo motors at the hips and knees must precisely alternate between flexion and extension to mimic this natural balance and energy transfer.

Similarly, abduction–adduction governs lateral movements in the frontal plane. Abduction, or moving the legs outward, helps tilt the skis and initiate a turn, while adduction, bringing the legs back together, stabilizes the skier through the curve. In robotic skiing, actuators at the hips adjust these movements to control ski edge angles, while ankle adjustments fine-tune edge contact for precision steering.

To execute smooth and stable carved turns, both humans and robots rely on a synergy of movements, where extension–flexion manages the vertical motion and abduction–adduction ensures lateral weight shifts. By integrating these mechanisms, humanoid ski robots can navigate slopes with increasing efficiency, closely imitating the biomechanics of human skiing.

These biomechanical actions allow robots to adjust their posture dynamically. Stability during skiing is a critical factor, with previous studies emphasizing the importance of maintaining the ZMP within the support polygon of the robot [5,20,22]. This principle holds regardless of whether the humanoid has an active torso joint [3] or not [4].

4.4.4. Ski Designs and Shapes

In the majority of the studies, researchers opted for carving skis rather than traditional straight skis, primarily due to their enhanced ability to initiate and maintain turns without the need for skidding [3–6,20–23,49,51,52,54–57,59]. In contrast, a smaller number of studies utilized downhill skis [6,48], typically designed for high-speed stability on steep slopes, or freestyle skis [58,60], which are more flexible and symmetrical to support tricks and jumps in terrain parks. These choices reflect the specific performance demands and movement patterns being investigated in each discipline: slalom, which demands quick, sharp turns, downhill racing, where speed and stability are important, and freestyle skiing, which emphasizes agility and maneuverability.

Traditional skis, widely used before the 1990s, had a straight, narrow profile requiring significant edge control and skidding to turn. In contrast, carving skis feature a deep sidecut, creating an hourglass shape that allows skiers to tip the ski on edge and carve clean arcs with minimal effort. This innovation began with the Elan SCX in 1994, which marked the shift toward modern performance-oriented skis [61].

The curvature of a ski (sidecut radius), defined by its tip, waist, and tail width, determines how tight or wide turns will be. A smaller sidecut radius (10–15 m) makes skis more agile, ideal for slalom racing, while a larger radius (30–50 m) provides stability, suited for high-speed downhill runs. Beyond side cuts, the ski's camber and rocker profiles also significantly influence performance. The camber refers to the gentle upward arch in the middle of the ski, ensuring even weight distribution and better edge grip on hard snow. In contrast, a rocker, a reverse camber where the tip and tail rise, improves floatation in deep snow and makes turn initiation smoother. Figure 8 provides a visual representation of a ski's sidecut, along with its camber and rocker profiles.

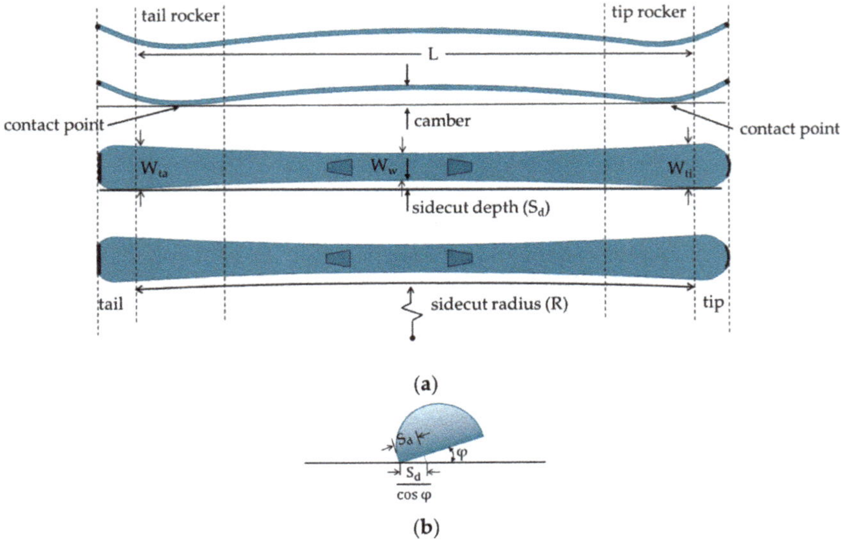

Figure 8. Illustration of ski geometry, showing sidecut, camber, and rocker: (**a**) top and side views; (**b**) front view of the ski flexed during a turn.

The ideal blend of camber and rocker varies by ski type. Slalom carving skis typically feature more camber and a sharper sidecut, maximizing precision and responsiveness on groomed runs. All-mountain skis, on the other hand, balance camber and rocker to offer versatility across different snow conditions, allowing for both stability and maneuverability.

The sidecut radius R (in meters) characterizes the curvature of a ski's edge and can be approximated by

$$R = \frac{L^2}{2(W_{ti} + W_{ta} - 2W_w)} \quad (1)$$

where L is the ski length and W_{ti}, W_w, and W_{ta} are the tip, waist, and tail widths, respectively. Assuming a circular arc, this can be expressed more compactly by defining the sidecut depth S_d as [62]:

$$S_d = \frac{W_{ti} + W_{ta} - 2W_w}{4} \Rightarrow R = \frac{L^2}{8\,S_d} \quad (2)$$

This formulation highlights the inverse relationship between the sidecut radius and the sidecut depth: A deeper sidecut (larger S_d) results in a smaller turning radius, which enhances the ski's ability to make tighter turns.

This shows that a deeper sidecut (larger S_d) leads to a smaller turning radius. In actual skiing, the effective turn radius (R_t) is further reduced by the ski's edge angle φ, according to

$$R_t = R\cos\varphi \quad (3)$$

4.4.5. Simulation, Experimentation, and Validation Techniques

In terms of software used in the studies, ski robots are tested in both simulated and real-world environments, each offering distinct advantages. Simulations provide controlled conditions, making it easier to test algorithms without the unpredictable variables of real snow. Various software tools, including Webots [5], MATLAB [20,48,49,56,57], Blender [49], Unity [48], and ADAMS [58,60], are used to model dynamics, stability, and autonomous control. Some researchers even develop custom simulators to better replicate skiing physics.

For example, Webots enable testing carving turns and navigation [5], while MATLAB's SimMechanics toolbox helps model skiing robot dynamics [49]. Blender assists with visualizing ski environments [49], and Unity is used for virtual reality applications in adaptive skiing [48].

Real-world testing, on the other hand, is essential for validating simulation results. Many experiments are conducted on actual ski slopes, though snow conditions can be unpredictable. To ensure consistency, some studies use artificial slopes, such as polyester-covered surfaces or paved tracks with roller skis [23]. Other setups, like Stewart platforms, simulate ground inclination changes, creating hybrid environments that blend physical hardware with virtual simulations [21,51]. Industrial robots are also used in laboratories to analyze ski deflection and binding release mechanisms under controlled conditions [50,57].

4.5. Limitations and Challenges of Skiing Experiments

Studies on skiing robots encounter several challenges that limit their accuracy and real-world applicability. One of the main difficulties is simulating real snow. Many researchers struggle to replicate the complex properties of snow [5], and even advanced simulators like Webots fail to provide a truly realistic environment. This makes it hard to verify algorithms under actual skiing conditions. Artificial test surfaces, like artificial grass, are often used in experiments, but they do not accurately mimic the friction properties of real snow [7]. This difference leads to discrepancies in how the robot behaves compared to real-world conditions.

Another limitation is the lack of simulation tools that can accurately model skiing robots, skis, and terrain in real time [5]. This gap has led some researchers to develop their own systems for testing. Additionally, some skiing robot simulations do not offer an easy way to control the robot's speed while in a sliding state, making it difficult to replicate actual skiing dynamics.

In experiments with physical skiing robots, there is also a narrow range of parameters that can result in successful turns [7]. These parameters, such as joint angles, are influenced by various factors, including speed, slope angle, and the robot's physical characteristics. This makes it challenging to find a consistent method for achieving reliable results.

The mechanical models used to simulate ski turns are another hurdle. These models often involve simplifications and approximations, which may not fully represent the actual movements made by skiers [23]. This can make the models difficult to use for skiers trying to improve their technique.

Skiing robots also face challenges related to sensor development [20]. The sensors required to accurately replicate skiing are still under development, making skiing a particularly difficult task for robots. Stability control is another issue. Some skiing robots cannot properly maintain stability, especially when handling uneven surfaces or maintaining the ZMP in all necessary axes [20].

Safety binding tests using robots also face simplifications in motion patterns that fail to capture the full complexity of real falls. In some experiments, there is a lack of standardized procedures, such as for defining pivot points in binding tests, which can affect the interpretability and comparability of results [50]. Even with accurate robots and calibrated sensors, laboratory setups can still introduce systematic errors, which are difficult to eliminate without complex in situ calibration.

In terms of deflection measurement, validating ski deflection sensors in dynamic conditions remains a challenge [57]. Previous methods, such as strain gauges, often lacked rigorous validation, and new optical systems for detecting curvature face limitations in marker distance and single sensor validation. Similarly, while prototypes for detecting torsional deflection show promise, their accuracy needs to be verified against a "gold standard" measurement, as laboratory simulations often fail to replicate real-world support conditions [56].

Lastly, friction is a critical factor in skiing, but many studies on passive skiing robots fail to consider it properly [54]. Without incorporating friction models, it is difficult to compare robot simulations with real-world skiing conditions. These challenges highlight the need for further research and development to bridge the gap between theoretical simulations and actual skiing performance.

A comparative analysis of experimental approaches highlights marked differences in robot performance between simulations and real-world scenarios. Robots in virtual environments often perform smoothly due to simplified physics, while real-world counterparts face instability and delayed responses caused by sensor noise, latency, and unpredictable terrain conditions. To address these discrepancies, future efforts should explore hybrid testing setups that integrate real hardware with simulated environments, such as motion platforms or augmented reality systems, to mimic varied terrain while incorporating real-time sensor feedback for adaptive control.

From a control systems standpoint, strategies like adaptive impedance control [63] and reinforcement learning-based feedback [64] show strong potential for handling dynamic interactions. These approaches have demonstrated effectiveness in enhancing control precision and mechanical resilience in related applications [65]. By combining such advanced control techniques with scalable test platforms, this review outlines a more effective and realistic direction for future skiing robotics research and development.

To accelerate progress in skiing robotics, key technical challenges must be prioritized. These include developing accurate, real-time models of snow-surface interaction and friction using sensor fusion and machine learning, enhancing environmental perception through robust, multi-modal sensor systems, and advancing adaptive control algorithms that adjust to dynamic terrain conditions. Equally important are the creation of realistic, open-access simulation platforms to support development and collaboration and the establishment of standardized testing protocols to ensure consistent evaluation. Addressing these challenges will not only advance skiing robotics but also benefit related fields like humanoid robotics and autonomous mobility.

4.6. Practical Applications of Robotics Related to Skiing and Future Development Paths

Beyond their role as technological demonstrators, skiing robots can provide a wide range of practical applications:

- Standardized ski equipment testing: robotic systems capable of replicating controlled skiing maneuvers and providing reproducible test conditions for evaluation of ski designs, boot stiffness, and binding safety mechanisms;
- Athlete training and performance analysis: skiing robots equipped with biomechanical modeling that can act as performance references for human skiers, providing real-time feedback on optimal movement patterns and supporting technique refinement;
- Rehabilitation and assistive technologies: Robotic skiing platforms can be adapted for rehabilitation purposes, helping athletes recovering from injuries or supporting training programs;
- Research in unstructured terrain locomotion: The challenges faced by skiing robots can provide valuable insights for other fields such as autonomous exploration robotics, disaster response systems, and planetary rover development.

To accelerate progress toward these applications, future research should prioritize:

- The integration of sensor fusion techniques (combining LiDAR, IMU, GNSS, and vision systems) to enhance perception and environmental modeling;

- The adoption of reinforcement learning and adaptive control frameworks for dynamic decision-making on variable terrains;
- Biomechanical optimization of joint actuation patterns, mimicking human skiing postures for better energy efficiency and stability;
- Customization of skiing robots and exoskeletons for rehabilitation exercises on dry slopes or indoor simulators;
- Combining robotics with virtual reality environments to create immersive and motivational therapy programs;
- Modular hardware design to enable rapid adaptation of robots to different skiing styles (e.g., alpine, freestyle, and slalom).

These targeted development paths will help transition skiing robotics from experimental prototypes to valuable platforms for sports science, equipment testing, and autonomous system development.

4.7. Limitations of the Study

Despite the systematic and comprehensive approach used in this review, the analysis revealed only limited new information that could significantly push the boundaries of current research on robotic applications in skiing. This result stems from several contributing factors.

First, the available body of literature remains limited in both volume and scope. With only 24 eligible studies identified, the field is still emerging, and much of the existing work converges around similar challenges, such as carving turn control, stability maintenance, and obstacle avoidance. Consequently, many studies reiterate known technical challenges, particularly difficulties in simulating realistic snow conditions, controlling stability via ZMP techniques, and replicating human biomechanics in dynamic skiing environments.

Second, the methodological diversity among the studies, while valuable for scope, made it difficult to synthesize findings into new generalizable theories or frameworks. Differences in experimental design, robot types, and evaluation metrics created a fragmented landscape rather than a cohesive, cumulative body of knowledge.

Third, the persistence of common limitations across multiple studies, including the overreliance on laboratory settings, inadequate modeling of ski–snow interaction, and the absence of standardized performance evaluation protocols, limited the capacity of the review to identify disruptive technological advancements or transformative approaches.

Moreover, while the integration of machine learning techniques is gaining traction, current applications remain in preliminary stages, often restricted to simulations without comprehensive real-world validation. This restricts their ability to generate actionable insights for immediate practical deployment.

In summary, the analysis reflects a field that is still in its exploratory phase, marked more by incremental progress and repeated challenges than by breakthrough innovations. While the review serves to map the existing terrain, identify recurring obstacles, and highlight promising directions, it ultimately confirms that robotic skiing research remains nascent and fragmented. Significant conceptual and technological leaps will be necessary before more profound and impactful insights can be realized.

Given the emerging nature of skiing robotics, the number of available studies remains limited. While the systematic methodology applied here is designed for larger datasets, this review serves as a necessary first step, mapping the current landscape and establishing a framework for future, more extensive analyses as the field expands.

5. Conclusions

This systematic review examined the current state of research on robotic applications in skiing, analyzing 24 eligible studies from diverse geographical locations and research institutions. The findings reveal several important trends and identify substantial opportunities for future development in this emerging field.

The research landscape is characterized by concentrated contributions from a few key countries (Slovenia, China, South Korea, and Japan) and institutions (notably the Jožef Stefan Institute and Kwansei Gakuin University), suggesting potential benefits from broader international collaboration and knowledge exchange. The temporal distribution of publications indicates steady but modest growth in research interest since 2009, with carving turn control, stability control, and obstacle avoidance emerging as dominant research aims.

A notable observation from this review is the strong preference for carving skis in robotic skiing applications, utilizing their geometric design to enable smooth turn execution without skidding. The predominant use of humanoid robots and lower-body mechanical platforms reflects attempts to replicate human biomechanics, particularly the extension–flexion and abduction–adduction movements essential for executing carved turns. Maintaining the ZMP within the support polygon has emerged as a critical stability control strategy across multiple studies.

This review identified significant challenges that currently limit the advancement of skiing robotics. The difficulty in accurately simulating snow properties and ski–snow interactions represents perhaps the most fundamental obstacle, leading to discrepancies between simulated performance and real-world behavior. The narrow parameter ranges for successful turns, limitations in sensor technology, and difficulties in stability control on variable terrain further complicate development efforts. Additionally, the lack of standardized testing methodologies hampers comparative analysis and benchmarking across different robotic systems.

Looking forward, several promising research directions emerge. The integration of advanced machine learning techniques, particularly deep learning for environmental perception and adaptive control, offers potential solutions for real-time terrain adaptation. Developing more sophisticated physical models of ski–snow interaction could bridge the gap between simulation and real-world performance. Furthermore, greater emphasis on interdisciplinary collaboration between robotics engineers, biomechanics experts, and skiing professionals could yield more practical and effective robotic systems.

The applications of skiing robotics extend beyond sports performance to include equipment testing, rehabilitation technology, and training aids. Robotic systems capable of precisely replicating human skiing movements offer unique opportunities for standardized equipment testing and biomechanical analysis that would be difficult to achieve with human subjects alone.

In conclusion, while robotics in skiing remains a specialized field with significant technical challenges, its continued development holds promise for advancing both the sport itself and broader applications in human–robot interaction, rehabilitation robotics, and sports science. Addressing the identified limitations through interdisciplinary collaboration, improved simulation technologies, and rigorous field testing will be essential for realizing the full potential of robotic applications in skiing.

Funding: This research received no external funding.

Data Availability Statement: The data presented in this study are available on request from the corresponding author.

Acknowledgments: During the preparation of this manuscript/study, the author used Claude 3.7 Sonnet for the purposes of text editing. The author has reviewed and edited the output and takes full responsibility for the content of this publication.

Conflicts of Interest: The author declares no conflicts of interest.

References

1. Wang, K.; Jing, Z.; Gang, L.; Yang, H.; Hu, F. Industrial automation and product quality: The role of robotic production transformation. *Appl. Econ.* **2024**, 1–16. [CrossRef]
2. Liu, S.; Liu, Z.; Daniel, A.; Thilak, D. AI with Robotics for leg support to skiers and snowboarders. *J. Intell. Fuzzy Syst.* **2021**, 1–13. [CrossRef]
3. Nemec, B.; Lahajnar, L. Control and navigation of the skiing robot. In Proceedings of the 2009 IEEE/RSJ International Conference on Intelligent Robots and Systems, IROS 2009, St. Louis, MO, USA, 11–15 October 2009; pp. 2321–2326.
4. Han, Y.H.; Jeon, H.J.; Cho, B.K. Development of a Humanoid Robot for the 2018 Ski Robot Challenge. *Int. J. Precis. Eng. Manuf* **2020**, *21*, 1309–1320. [CrossRef]
5. Kim, S.H.; Lee, B.; Hong, Y.D. Stability control and turning algorithm of an alpine skiing robot. *Sensors* **2019**, *19*, 3664. [CrossRef]
6. Iverach-Brereton, C.; Postnikoff, B.; Baltes, J.; Hosseinmemar, A. Active balancing and turning for alpine skiing robots. *Knowl Eng. Rev.* **2016**, *32*, e6. [CrossRef]
7. Yoneyama, T.; Kagawa, H.; Unemoto, M.; Iizuka, T.; Scott, N.W. A ski robot system for qualitative modelling of the carved turn. *Sports Eng.* **2009**, *11*, 131–141. [CrossRef]
8. Erdmann, W.S. Problems of Sport Biomechanics and Robotics. *Int. J. Adv. Robot. Syst.* **2013**, *10*, 123. [CrossRef]
9. Zheng, Y.F.; Wang, H.; Li, S.; Liu, Y.; Orin, D.; Sohn, K.; Jun, Y.; Oh, P. Humanoid robots walking on grass, sands and rocks. In Proceedings of the IEEE Conference on Technologies for Practical Robot Applications, TePRA, Woburn, MA, USA 22–23 April 2013.
10. Yin, K.; Gao, Y.; Gao, F.; Chen, X.; Zhao, Y.; Xiao, Y.; Sun, Q.; Sun, J. Mechanism Design and Motion Planning of a Hexapod Curling Robot Exhibited During the Beijing 2022 Winter Olympics Games. *Engineering* **2024**, *35*, 15–31. [CrossRef]
11. Lin, H.-I.; Yu, Z.; Huang, Y.-C. Ball Tracking and Trajectory Prediction for Table-Tennis Robots. *Sensors* **2020**, *20*, 333. [CrossRef]
12. Riener, R.; Rabezzana, L.; Zimmermann, Y. Do robots outperform humans in human-centered domains? *Front. Robot. AI* **2023** *10*, 1223946. [CrossRef]
13. Boboc, R. Enhancing Efficiency and Performance: Robotic Applications in Skiing. In Proceedings of the 5th IFSA Winter Conference on Automation, Robotics & Communications for Industry 4.0/5.0 (ARCI' 2025), Granada, Spain, 19–21 February 2025 pp. 7–12.
14. Siegel, J.; Morris, D. Robotics, Automation, and the Future of Sports. In *21st Century Sports: How Technologies Will Change Sports in the Digital Age*; Schmidt, S.L., Ed.; Springer International Publishing: Cham, Switzerland, 2024; pp. 67–85.
15. Hu, L.; Liu, C.; Jiang, W.; He, D.; Qu, Q.; Wang, Z.; Ying, C. Application of gaming robot based on gait recognition algorithm in sports training and assistance system. *Entertain. Comput.* **2025**, *52*, 100763. [CrossRef]
16. Iverach-Brereton, C.; Baltes, J.; Anderson, J.; Winton, A.; Carrier, D. Gait design for an ice skating humanoid robot. *Robot. Auton Syst.* **2014**, *62*, 306–318. [CrossRef]
17. Javier Lopez Frias, F.; Luis Pérez Triviño, J. Will robots ever play sports? *Sport Ethics Philos.* **2016**, *10*, 67–82. [CrossRef]
18. Trenchard, H. When Robots Get Bored and Invent Team Sports: A More Suitable Test than the Turing Test? *Information* **2018**, *9* 118. [CrossRef]
19. Liu, J. Application of entertainment and fitness robots based on game interaction in sports training data analysis. *Entertain Comput.* **2025**, *52*, 100837. [CrossRef]
20. Park, C.; Kim, B.; Kim, Y.; Eum, S.; Song, H.; Yoon, D.; Moon, J.; Han, J. Carved Turn Control with Gate Vision Recognition of a Humanoid Robot for Giant Slalom Skiing on Ski Slopes. *Sensors* **2022**, *22*, 816. [CrossRef]
21. Petrič, T.; Gams, A.; Babič, J.; Žlajpah, L. Reflexive stability control framework for humanoid robots. *Auton. Robot.* **2013**, *34*, 347–361. [CrossRef]
22. Wu, Z.; Ye, J.; Wang, X.; Li, F. Control of Ski Robot Based on Deep Reinforcement Learning. In Proceedings of the Conference Digest—2021 International Conference on Security, Pattern Analysis, and Cybernetics, SPAC 2021, Chengdu, China, 18–20 June 2021; pp. 211–215.
23. Kono, K.; Saga, N. Development of a passive turn type skiing robot with variable height mechanism of gravitational center. *J. Robot. Mechatron.* **2012**, *24*, 372–378. [CrossRef]
24. Yang, L.; Yin, Y.; Wang, Z.; Wang, L.; Gao, F.; Chen, X.; Gao, H. Design and Control of a Novel Six-Legged Robot for Flat, Downhill, and Uphill Skiing. In *Lecture Notes in Computer Science (Including Subseries Lecture Notes in Artificial Intelligence and Lecture Notes in Bioinformatics)*; Springer: Singapore, 2023; pp. 469–480.

25. García, J.M.; Duarte, F.G. Mobile rolling robots designed to overcome obstacles: A review. *Forces Mech.* **2024**, *16*, 100283. [CrossRef]
26. Ju, F.; Wang, Y.; Xie, B.; Mi, Y.; Zhao, M.; Cao, J. The Use of Sports Rehabilitation Robotics to Assist in the Recovery of Physical Abilities in Elderly Patients with Degenerative Diseases: A Literature Review. *Healthcare* **2023**, *11*, 326. [CrossRef]
27. Wu, Z.; Zheng, K.; Ding, Z.; Gao, H. A survey on legged robots: Advances, technologies and applications. *Eng. Appl. Artif. Intell.* **2024**, *138*, 109418. [CrossRef]
28. Yu, Z.; Duan, A.; Zhu, Z.; Zhang, W. Biomimetic soft-legged robotic locomotion, interactions and transitions in terrestrial, aquatic and multiple environments. *Sustain. Mater. Technol.* **2024**, *40*, e00930. [CrossRef]
29. Bakırcıoğlu, V.; Çabuk, N.; Jond, H.B.; Kalyoncu, M. Optimization-driven design and experimental validation of a hydraulic robot leg mechanism. *Measurement* **2025**, *250*, 117096. [CrossRef]
30. Chambers, V.; Hobbs, B.; Gaither, W.; Thé, Z.; Zhou, A.; Karakasis, C.; Artemiadis, P. The Variable Stiffness Treadmill 2: Development and Validation of a Unique Tool to Investigate Locomotion on Compliant Terrains. *J. Mech. Robot.* **2025**, *17*, 031005. [CrossRef]
31. Chen, Z.; Chen, J.; Jung, S.; Kim, H.-Y.; Lo Preti, M.; Laschi, C.; Ren, Z.; Sitti, M.; Full, R.J.; Yang, G.-Z. Bioinspired and biohybrid soft robots: Principles and emerging technologies. *Matter* **2025**, *8*, 102045. [CrossRef]
32. Cui, J.; Wang, P.; Sun, T.; Ma, S.; Liu, S.; Kang, R.; Guo, F. Design and experiments of a novel quadruped robot with tensegrity legs. *Mech. Mach. Theory* **2022**, *171*, 104781. [CrossRef]
33. Marcus, S.D.; Shapiro, A.; Giladi, C. Enhancing Quadruped Robot Walking on Unstructured Terrains: A Combination of Stable Blind Gait and Deep Reinforcement Learning. *Electronics* **2025**, *14*, 1431. [CrossRef]
34. Wang, L.; Yin, Y.; Yang, L.; Gao, F.; Chen, X. Development and Implementation of a Six-Legged Skiing Robot for Cross-Country Skiing Techniques. In *Lecture Notes in Computer Science (Including Subseries Lecture Notes in Artificial Intelligence and Lecture Notes in Bioinformatics)*; Springer: Singapore, 2025; pp. 185–198.
35. Afzal, N.; Rehman, M.u.; Seneviratne, L.; Hussain, I. The Convergence of AI and animal-inspired robots for ecological conservation. *Ecol. Inform.* **2025**, *85*, 102950. [CrossRef]
36. Nguyen, V.C.; Kim, S.H. A novel fixed-time prescribed performance sliding mode control for uncertain wheeled mobile robots. *Sci. Rep.* **2025**, *15*, 5340. [CrossRef]
37. Xue, Y.; Wang, L.; He, B.; Zhao, Y.; Wang, Y.; Li, L. Research on Environmental Adaptability of Force–Position Hybrid Control for Quadruped Robots Based on Model Predictive Control. *Electronics* **2025**, *14*, 1604. [CrossRef]
38. Khan, R.A.I.; Zhang, C.; Deng, Z.; Zhang, A.; Pan, Y.; Zhao, X.; Shang, H.; Li, R. Multi-Agent Reinforcement Learning Tracking Control of a Bionic Wheel-Legged Quadruped. *Machines* **2024**, *12*, 902. [CrossRef]
39. Jamali, A.; Labadin, A.A.; Jamaludin, M.S. Investigation of Hybrid Mechanism for Knee Exoskeleton Rehabilitation Device. In Proceedings of the IEEE International Conference on Smart Instrumentation, Measurement and Applications, ICSIMA, Bandung, Indonesia, 30–31 July 2024; pp. 109–113.
40. Ma, C.; Jiang, K.; Zhao, Q.; Ni, D.; Zhang, J. Innovative Applications of Multimodal Sensing Technology in Sports Rehabilitation Assessment and Training. *Scalable Comput.* **2025**, *26*, 855–863. [CrossRef]
41. Marchesi, G.; Ricaldone, E.; De Luca, A.; Torre, K.; Quinland, E.; Bellitto, A.; Squeri, V.; Massone, A.; Casadio, M.; Canessa, A. A robot-based assessment of trunk control in Spinal Cord Injured athletes. In Proceedings of the IEEE RAS and EMBS International Conference on Biomedical Robotics and Biomechatronics, New York, NY, USA, 29 November–1 December 2020; pp. 497–502.
42. Li, X.-A.; Zhang, D.; Jia, X.; Han, X.; Ning, G.; Li, Q. A Bayesian learning approach for dynamic parameter identification and its applications in industrial robotic systems. *Comput. Methods Appl. Mech. Eng.* **2025**, *441*, 117951. [CrossRef]
43. Zhang, D.; Liang, H.; Li, X.-A.; Jia, X.; Wang, F. Kinematic calibration of industrial robot using Bayesian modeling framework. *Reliab. Eng. Syst. Saf.* **2025**, *253*, 110543. [CrossRef]
44. Page, M.J.; McKenzie, J.E.; Bossuyt, P.M.; Boutron, I.; Hoffmann, T.C.; Mulrow, C.D.; Shamseer, L.; Tetzlaff, J.M.; Akl, E.A.; Brennan, S.E.; et al. The PRISMA 2020 statement: An updated guideline for reporting systematic reviews. *BMJ* **2021**, *372*, n71. [CrossRef]
45. Singh, J. Critical appraisal skills programme. *J. Pharmacol. Pharmacother.* **2013**, *4*, 76–77. [CrossRef]
46. Sterne, J.A.C.; Savović, J.; Page, M.J.; Elbers, R.G.; Blencowe, N.S.; Boutron, I.; Cates, C.J.; Cheng, H.-Y.; Corbett, M.S.; Eldridge, S.M.; et al. RoB 2: A revised tool for assessing risk of bias in randomised trials. *BMJ* **2019**, *366*, l4898. [CrossRef]
47. Hasan, M.K. P2P Motion: A Motion Planning Algorithm for Cyclic Synchronous Position Based Motion Control Systems. In Proceedings of the International Conference on Electronics, Communications and Information Technology, ICECIT 2021, Khulna, Bangladesh, 14–16 September 2021.
48. Houda, T.; Beji, L.; Amouri, A.; Mallem, M. Dynamic behavior of an interactive XY—6 DoF simulator for people with reduced mobility. In Proceedings of the CCTA 2020—4th IEEE Conference on Control Technology and Applications, Montreal, QC, Canada, 24–26 August 2020; pp. 522–527.
49. Lahajnar, L.; Kos, A.; Nemec, B. Skiing robot—Design, control, and navigation in unstructured environment. *Robotica* **2009**, *27*, 567–577. [CrossRef]

50. Nimmervoll, F.; Eckerstorfer, R.; Braumann, J.; Petutschnigg, A.; Sternad, B. Method to Investigate Multi-Axis Release Action of Ski Safety Bindings: A New Approach for Testing in Research and Development. *Front. Sports Act. Living* **2021**, *3*, 585775. [CrossRef]
51. Petrič, T.; Nemec, B.; Babič, J.; Žlajpah, L. Multilayer control of skiing robot. In Proceedings of the IEEE International Conference on Intelligent Robots and Systems, San Francisco, CA, USA, 25–30 September 2011; pp. 4832–4837.
52. Petrič, T.; Peternel, L.; Gams, A.; Nemec, B.; Žlajpah, L. Navigation methods for the skiing robot. *Int. J. Humanoid Robot.* **2013**, *10*, 1350029. [CrossRef]
53. Rossi, S.; Cappa, P.; Patanè, F. Experimental measurement of the ski boot stiffness in sagittal and frontal planes. In Proceedings of the ASME International Mechanical Engineering Congress and Exposition, Lake Buena Vista, FL, USA, 13–19 November 2009; pp. 183–186.
54. Saga, T.; Saga, N. Alpine skiing robot using a passive turn with variable mechanism. *Appl. Sci.* **2018**, *8*, 2643. [CrossRef]
55. Saga, N.; Kono, K. Experimental study on alpine skiing turn using passive skiing robot. In *Lecture Notes in Computer Science (Including Subseries Lecture Notes in Artificial Intelligence and Lecture Notes in Bioinformatics)*; Springer: Berlin/Heidelberg, Germany, 2009; pp. 1044–1050.
56. Thorwartl, C.; Tschepp, A.; Holzer, H.; Stöggl, T. Assessing Torsional Deflection of Alpine Skis A Proof of Concept Utilizing Printed Electronics in a Laboratory Setting. In Proceedings of the 2023 IEEE SENSORS, Vienna, Austria, 29 October–1 November 2023; pp. 1–4.
57. Thorwartl, C.; Kröll, J.; Tschepp, A.; Holzer, H.; Teufl, W.; Stöggl, T. Validation of a Sensor-Based Dynamic Ski Deflection Measurement in the Lab and Proof-of-Concept Field Investigation. *Sensors* **2022**, *22*, 5768. [CrossRef] [PubMed]
58. Wu, X.; Yi, J. Modeling and kinematics simulation of freestyle skiing robot. *Acta Tech. CSAV (Ceskoslovensk Akad. Ved)* **2017**, *62*, 321–334.
59. Yang, L.; Wang, Z.; Wang, L.; Fu, Y.; Yin, Y.; Gao, F.; Gao, H. Obstacle Avoidance Strategy for a Novel Skiing Robot in Unknown Snow Environments. *IEEE Robot. Autom. Lett.* **2024**, *9*, 3203–3210. [CrossRef]
60. Zhenyu, W.; Baichun, L.; Tianbiao, Y.; Wanshan, W.; Jian, L. Research on key technology of freestyle skiing robot in the aid slippery stage. In Proceedings of the IEEE International Conference on Industrial Technology, Seville, Spain, 17–19 March 2015; pp. 3202–3206.
61. Truong, J.; Bulota, M.; Desbiens, A.L. Historical Trends in Alpine Ski Design: How Skis Have Evolved Over the Past Century. *Proceedings* **2020**, *49*, 135. [CrossRef]
62. Lind, D.A.; Sanders, S.P. Equipment: Properties and Performance. In *The Physics of Skiing: Skiing at the Triple Point*; Springer: New York, NY, USA, 2004; pp. 44–77.
63. Ma, J.; Chen, H.; Liu, X.; Yang, Y.; Huang, D. Adaptive Impedance Control of a Human–Robotic System Based on Motion Intention Estimation and Output Constraints. *Appl. Sci.* **2025**, *15*, 1271. [CrossRef]
64. Jeon, H.; Kim, D.-W.; Kang, B.-Y. Deep reinforcement learning for cooperative robots based on adaptive sentiment feedback. *Expert Syst. Appl.* **2024**, *243*, 121198. [CrossRef]
65. Hoseinnezhad, R. A Comprehensive Review of Deep Learning Techniques in Mobile Robot Path Planning: Categorization and Analysis. *Appl. Sci.* **2025**, *15*, 2179. [CrossRef]

machines

Article

A Connective Framework for Social Collaborative Robotic System

Syed Osama Bin Islam * and Waqas Akbar Lughmani

Department of Mechanical Engineering, Capital University of Science and Technology (CUST), Islamabad Expressway, Zone 5, Islamabad 45750, Pakistan
* Correspondence: dme163001@cust.pk

Abstract: Social intelligence in robotics appeared quite recently in the field of artificial intelligence (AI) and robotics. It is becoming increasingly evident that social and interaction skills are essentially required in any application where robots need to interact with humans. While the workspaces have transformed into fully shared spaces for performing collaborative tasks, human–robot collaboration (HRC) poses many challenges to the nature of interactions and social behavior among the collaborators. The complex dynamic environment coupled with uncertainty, anomaly, and threats raises questions about the safety and security of the cyber-physical production system (CPPS) in which HRC is involved. Interactions in the social sphere include both physical and psychological safety issues. In this work, we proposed a connective framework that can quickly respond to changing physical and psychological safety state of a CPPS. The first layer executes the production plan and monitors the changes through sensors. The second layer evaluates the situations in terms of their severity as anxiety by applying a quantification method that obtains support from a knowledge base. The third layer responds to the situations through the optimal allocation of resources. The fourth layer decides on the actions to mitigate the anxiety through the allocated resources suggested by the optimization layer. Experimental validation of the proposed method was performed on industrial case studies involving HRC. The results demonstrated that the proposed method improves the decision-making of a CPPS experiencing complex situations, ensures physical safety, and effectively enhances the productivity of the human–robot team by leveraging psychological comfort.

Keywords: human–robot collaboration; cyber-physical production system; social safety; optimization; artificial intelligence

1. Introduction

Conventionally, robots were used with the capabilities of nominal sensors and intelligence in the industry. The tasks assigned to them were normally repetitive in nature [1]. However, recent developments in the industry have spawned a wide range of robots. The up-gradations are common in services [2,3], exploration and rescue [4,5], and therapy operations [6,7]. The advent of human and robot interaction in the above-mentioned fields has given rise to the specialized field of human–robot interaction (HRI) [8]. The field of HRI is now transforming into social HRI and is of particular importance. The interactions in this field include cognitive, social, and emotional activities with the robots [9]. Moreover, a new term of human–machine collaboration (HMC) arises in this context, and human–robot collaboration (HRC) specifically for robots. Chandrasekaran et al. [10] described HRC as a measure for improved task performance of the robot and reduced tasks for humans. A shared space is used by humans and robots while performing collaborative tasks, and it is expected that they attain a common objective while conforming to the rules of social interaction. These robots perform joint actions, obey rules of social interaction (such as proxemics) and still act efficiently and legibly [11]. Since these interactions take place continuously during collaborative human and robot tasks, they place strong demands on collaborators' personal and environmental safety. In addition to physical contact, the control system of the human–robot collaboration system must cater to uncertainties and ensure stability. Mead et al. [12] identified three types of requirements for HRI commonly used by computer models of proxemics:

- Physical requirements, based on collaborators' distance and orientation;
- Psychological requirements, based on the collaborators' relationship;
- Psychophysiological requirements based on the sensory experience through social stimuli.

The broad categories of safety for the robotic collaborative system in the social domain are physical and psychological safety. The earlier provides safety from physical hazards, whereas the latter is from psychological discomforts such as close interaction with machines, monotonous operations, or deviations from the task.

Different physical safety protocols for HRC-based CPS were presented by authors [13–16], whose activation is dependent on the proximity between the cobot and the human operator. Collision avoidance is a common solution to provide physical safety that includes avoiding unwanted contact with people or environmental obstacles [17]. These techniques depend on the measurement of the distance in-between the robot and the obstacle [16,18]. Motion planning techniques are one of the key strategies to estimate collision-free trajectories such as configuration × time–space method [19]; collision-free vertices [20,21]; representation of objects as spheres; and exploring collision-free path [22], potential field method [23], and virtual spring and damping methods [24]. Unfortunately, collision avoidance can fail because of the sensors and robotic movement limitations, as sometimes human actions are quicker than robotic actions. It is nevertheless feasible to sense the bodily collision and counteract it [25–27], which allows eradicating the robot from the contact area. In such circumstances, robots can use variable stiffness actuation [28,29] while effectively controlled [30,31], and lightweight robots with compliant joints [32] may be used to lessen the impact forces on the contact. Lately, a multi-layered neural network method involving dynamics of the manipulator joints (measurement of torque through sensors and intrinsic joint positions through kinematics) has been used [33] to discover the location of the collision on the robot (collided link). While accidents due to sudden contact between humans and robots may be restricted by designing lightweight/compliant mechanical manipulators [30] and collision detection/response strategies [27], collision avoidance in complex, unpredictable, and dynamic environments is still mainly dependent on the employment of exteroceptive sensors.

From the perspective of HRI, psychological safety means ensuring interactions that do not cause undue stress and discomfort in the long run. Therefore, a robot that can sense worker fatigue with whom it works is able to take essential safeguards to avoid accidents. An illustration of it was demonstrated in [34], a human–computer integration where biosensors were mounted on the operator, and physiological signals (such as anxiety) were measured and transmitted to the robot. Researchers are required to build robots that must respond to emotional information received from the operator, also known as "affective computing" [35]. In [36–38], measurements of cardiac activity, skin conductivity, respiration, and eye muscle activity in conjunction with fuzzy inference tools were used to change the robot's trajectory for safe HRC. Recently, Chadalavada et al. [39] demonstrated how a robot (automated forklift) could show its intentions using spatial augmented reality (SAR) so that humans can visually understand the robot's navigational intent and feel safe next to it.

It seems now essential that methods ensuring both physical and psychological safety be developed for HRC. Lasota et al. [40] introduced a distinct concept of combining both physical and psychological safety for safe HRC. By measuring the distance between the robot and the human in real-time, they precisely control the speed of the robot at low separation distances for collision avoidance and stress relief. In another paper [14], the same authors established via quantitative metrics that human-conscious motion planning leads to more effective HRC. Lately, Dragan et al. [41] described that psychological safety is ensured by legibility; the operator may feel more comfortable if they can judge the robot's intention by its motion.

Humans, robots, and machines lately work in the ambit of cyber-physical systems (CPS) under the umbrella of Industry 4.0. They are smart systems that contain both physical and computational elements. The concept involves decision-making through real-time data evaluation collected from interconnected sensors [42]. Monostori [43] proposed overall automation of these layers for production systems interconnecting necessary physical elements such as machines, robots, sensors, and conveyors through computer science and termed it a cyber-physical production system (CPPS). Green et al. [44] highlighted two important components of an active system of collaboration between humans and robots. First, the adjustable autonomy of the robotic system is an essential part of an effective collaboration that enhances productivity. Second, awareness of situations, or knowing what is happening in the robot workspace, is also essential for collaboration. The effectiveness of HRC depends on the effective monitoring of human and environmental actions and the use of AI to predict actions and states of mind to which humans may have contributed to the task. Lemaignan et al. [11] emphasized the architecture of the decision layer of social robots. The paper is an attempt to characterize the challenges and present a series of critical decision problems that must be solved for cognitive robots to successfully share space and tasks with humans. In this context, the authors identified logical reasoning, perception-based situational assessment, affordability analysis, representation and acquisition of knowledge-based models for involved actors (humans and robots, etc.), multimodal dialogue, human-conscious mission planning, and HRC task planning.

As the industry is transforming from automation to intelligence, a need is felt to extend the concept of psychological safety from humans to the CPSs. Psychological safety is equally essential for an intelligent CPS as physical safety is perceived [45]. Although interactions among human operators, physical and computational layers of CPS can be quite demanding in terms of cognitive resources, the psychological aspects of safety are not extensively taken into account by existing systems. However, there is no connective framework that assesses and counters both physical and psychological issues of a CPPS in a social domain. Flexible CPPS in this regard is required to counter the physical and psychological uncertainties by defining contingencies. This indicates the requirement for an efficient and reliable framework for next-generation CPPS:

- A framework that can quantify both the physical and psychological safety of the CPPS;
- A framework that can assess the CPPS's current state and provide a thinking base accordingly to make it flexible and safe;
- A framework that can decide, optimize and control based on the situational assessment.

Our research focused on addressing both physical and psychological issues faced by a CPPS. We proposed a layered framework for knowledge-based decision-making in a CPS. The CPS, at the core, performs the desired operations through the interactions between its physical component (PC), computer component (CC), and human component (HC). A situational assessment layer is proposed above the central layer to assess the anxiety of the situations faced. Calculation of the matching score is suggested for the relevance of each situation's anxiety to each resource, and we named it an "anxiety factor". The third layer is the resource optimization layer, above the situational assessment layer. This layer optimizes the allocation of available resources through an optimization algorithm using the evaluated matching score, the objective function, and the defined constraints. The last is the logic-based decision-making layer. This layer embeds predefined logic to decide on complex situations, thereby tasking different resources using the experts' knowledge, evaluated optimization, and calculated anxieties. The logic remains specific to each identical case/situation embedded in a CPS scenario. The proposed framework is validated through experimental case studies facing several situations.

2. Methodology

The connective framework and an overview of its execution are shown in Figure 1. We propose a knowledge-based modular software system where different modules represent different layers of the proposed framework. However, we do not propose the number of modules to be fixed and may vary as per the requirement of a particular case. The data of all the layers are stored in a central database.

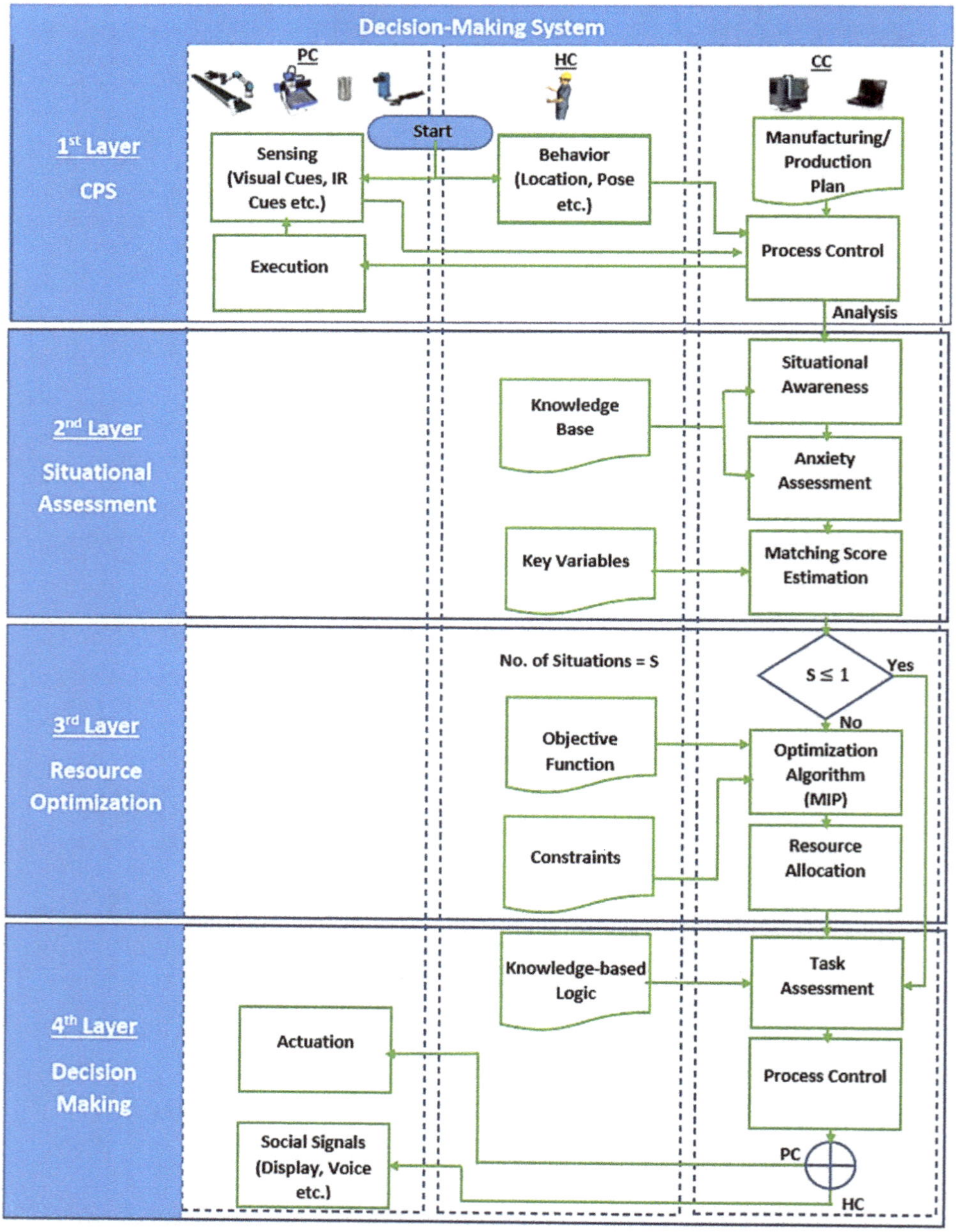

Figure 1. Decision-making connective framework for CPPS.

The basic modules are the main module for the sensing and process control layer of the CPS, the anxiety module for the situational assessment layer, the optimization module for the resource optimization layer, and the decision-making module for the decision-making layer. The general connectivity of the modules is shown in Figure 2.

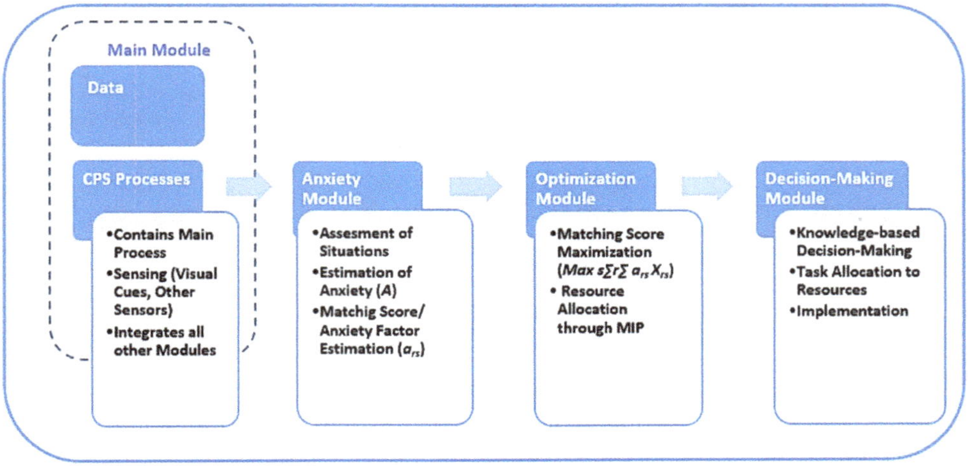

Figure 2. Module-based software implementation for decision-making.

If multiple changes are faced, the matching score is ascertained for each situation through the anxiety module. Accordingly, the situations with the higher matching score (anxiety factor) are assigned the best possible resources through the optimization module. The contingencies to handle the braved situations are looked after by the decision-making module. In this context, outputs from the anxiety and the optimization modules are given to the start of the decision-making module, which decides on the assignment of resources to the tasks based on the defined logic. The term anxiety and the anxiety factor are defined, and the method for their estimation is stated in this paper.

2.1. *Anxiety of Cyber-Physical Systems*

Anxiety is the human body's natural response to stress. It is a feeling of fear or apprehension about what is to come, also recognized as the unpleasant state when an expectation is not achieved due to any stressful, dangerous, or unfamiliar situation. The central task of the overall system in the brain is to compare, quite generally, actual with expected stimuli. We wanted to elaborate more on the term that should not be confused with risk. Risk is based on hazard, whereas "anxiety can be defined as an urge to perform any job either to avoid a hazard or to do a righteous job". When multiple situations are confronted, all generate anxiety, but there would be limited resources to handle each. Therefore, we defined anxiety to make it scalable for CPSs. There may be different levels of anxiety generated by situations. The module is initialized by the results of Ishikawa, which assigns an index to each anticipated situation. After initial indexing is performed, a novel and intelligent technique based on medical knowledge categorizes situations into different anxiety types. Each type relates to a particular level of severity.

2.1.1. Categorization of Anxiety

The criterion for categorizing and indexing anxiety in different situations faced by the CPS is defined in Table 1. The categories are named concerning the characteristics matched with the medical anxieties. Individual severity depends on the category and the repeatability of the situation. It can be said that the "Quantification" of psychological safety for the CPS is being performed. Anxiety for a situation is the sum of the lowest severity limit and the index "I" estimated by Ishikawa. The value of I ranges from 0 to 100. The detailed procedure for calculating index I is explained in [46].

Table 1. Anxiety Categories.

Level	Name	Description	Severity	Equation	Detailed Description
1	Panic	Emergency	>80 to 100	$P = 80 + I \times 20/100$	Medical: situation to which affected person is not accustomed and has a serious impact. CPS: a severe unknown incident identified by impact, action be taken by all stakeholders; however, it must be handled by HC, the most intelligent resource.
2	Post traumatic	Trauma/ Fear	>60 to 80	$T = 60 + I \times 20/100$	Medical: serious incident of the past, affected person remains under constant stress that it may occur again. CPS: assessment criterion is emergency declaration by HC on occurrence. Complete system must stop, and observations be rectified by the operator. After certain repetitions without any damage, the same may be transferred to specific phobia.
3	Agora-phobia/ Specific-Phobia	Known Specific Situation	>20 to 60	$K = 20 + I \times 40/100$	Medical: specific situation that is critical but known may be relieved based on available knowledge. CPS: known situations previously defined and not considered in panic/obsession/social norms. They may be transferred to the obsession or post-traumatic category on confirmation of false alarm or emergency.
4	General Anxiety	Intended Situation	20	$G = 20$	Medical: day-to-day routine situation for which refined solutions exist. CPS: ideal/intended situation. Act as a reference to categorize other situations.
5	Social norms	Etiquettes	>0 to 20	$N = 0 + I \times 20/100$	Medical: a communal problem not liked by humans. CPS: warnings that may not affect the current scenario, however, may affect the performance at later stage.
6	Obsession	False	0	$O = 0$	Medical: repetitive thought that leads to ritual or a false alarm. CPS: previously declared false alarms through Ishikawa, and a null value is assigned. Repetition and continuity are indicators; however, confirmation from HC is required.

Total severity is also calculated for a particular instance, and an alarm is raised if it crosses a certain limit. It is the sum of all current emerging situations in a single iteration, whose severities are estimated.

$$Total\ severity = G + P + O + T + K + N \quad (1)$$

2.1.2. Anxiety Factor Calculation

The anxiety factor, i.e., the matching score, is then calculated through a mathematical model dependent on the category and the variables related to tasks/resources. The procedure is repeated at each iteration and the situations faced are analyzed. The anxiety factor for a corresponding resource and a situation is calculated as:

$$a_{rs} = Q_{rs}\left[A_{rs} + (p_{rs} - t_{rs})\right] \quad (2)$$

Anxiety "A" is the parameter defined by the category of the particular situation, which is G, P, O, etc., and calculated as explained in Table 1. Its value ranges from 0 to 100. "p" is the preference variable and is the ascending order of anxiety level sorted for different situations; it can also be referred to as priority index number. "t" is the task variable; it defines which task can better be performed by which resource. The prior resource is assigned a lower value and subsequently ascending value for lower priorities. The value of t depends upon the number of resources, e.g., if there are two resources, then $t = 0, 1$. "Q" is the resource suitability variable; it has the value "1" if a resource is suitable, and "0" if not suitable. In order to simplify the mathematical notation of the model formulation, we defined the indices for resources "r" and situations "s".

The expression for "a" can be written as:

$a_{rs} \in [0,100 + S]$ for all resources $r \in R$ and situations $s \in S$;

$r \in R$: index and set of resources;

$s \in S$: index and set of situations.

S is the total no of situations. As the value of p ranges from 1 to S, hence the range of the anxiety factor is from 0 to 100 + S.

The anxiety factor "a", is the value of anxiety calculated through the above variables for a particular situation tackled by a specific resource.

2.2. *Layers of the Decision-Making System*

The framework is explained layer by layer. The CPs layer is the main layer that controls all physical and human elements. The physical component involves machines, robots, conveyors, sensors, display/output devices, input devices, etc. The main role of the layer is to execute the intended process for which the production plan is uploaded. It also looks for changes at every cycle of the operation. For this, various sensing techniques, such as proxemics and visual, physiological, or social cues, etc., may be used. Different situations are expected to affect the desired output are registered.

The situational assessment layer assess confronted situations for priority index (anxiety) by making use of the HC's knowledge base. First, the index of expected situations is calculated with the aid of Ishikawa analysis, as proposed in Section 2.1. The category of each situation is then identified. Based on the category, the matching score (anxiety factor) for all expected situations is calculated, which defines the resource suitability to the particular situation. On the initialization of the process, the layer first becomes aware of the situations that are detected by the CPS layer. The layer then links the matching scores to the related situations and re-estimates if the category of the situation changed by the HC during the operation.

The resource optimization layer optimizes the allocation of resources through an optimization algorithm. The module checks the number of situations; if there is a single situation, the program directly moves to the decision-making layer, and if there are multiple situations, the program moves to the optimization algorithm. The algorithm employs mixed-integer programming (MIP) technique making use of the matching score, objective function, and the defined constraints. The Gurobi Optimizer [47] is one of the state-of-the-art solvers for mathematical optimization problems. The MIP model of the stated problem is implemented in the Gurobi Optimizer. The overall technique addresses all the situations sequentially in terms of priority.

The decision-making module encompasses the logic defined by experts to handle the braved situations and decides on the assignment of resources to tackle them. The main consideration for the assignment of resources is the allocation recommendation by the optimization module; however, the implementation is carried out by the decision-making module. Commands are then given to the physical resources, which could be a human operator, a cobot, a machine, etc., depending on the ascertained tasks. The human component is also given cautions through social signals on observation of social norms and obsessions.

The implementation of the proposed method includes two types of actions, one to be taken before activating the system and others are happening in real-time during the process cycle. Both the pre-process and in-process steps concerning each layer are shown in Table 2.

Table 2. Process for decision-making CPPS.

Ser	Layer	Pre-Process Formalities	Process Pseudo-Code
1	**CPS**	Prepare Manufacturing/ Production plan	
2		Register expected situations	
3			Start
4			Upload Execution Plan
5			Initialize
6			Check state of human component
7			Check state of physical components
8			While no uncertainty
9			Execute the process
10			If uncertainty
11			Go to situational assessment layer
12			Else go to initialize
13			End
14	**Situational Assessment**	Index the defined situations' anxieties through Ishikawa method	
15		Categorize situations into the type of anxiety	
16		Assign weights to key variables for estimation of matching score	
17		Calculate the matching score (anxiety factor) with respect to resources	
18			Initialize
19			Upload the matching scores
20			Check the emerged situations
21			Check matching score of emerged situations
22			If category changed by operator
23			Re-designate the anxiety category of the situation
24			Re-estimate of matching score in case category is changed
25			Move to Resource Optimization Layer
26	**Resource Optimization**	Defining the optimization criteria for allocation of resources; the objective function and the constraints, MIP in our case	
27			Initialize
28			Case 1: Single situation
29			Move to Decision-Making Layer
30			
31			Case 2: Multiple situations
32			Allocate resource to situations through MIP
33			Move to Decision-Making Layer
34	**Decision Making**	Define and design the logic for each situation with inputs from experts for anxiety mitigation strategy	
35			Initialize
36			Upload resource allocation data
37			Check situations
38			Check resource allocation to situations
39			Assign task to available resources based on allocation and logic
40			Execute the task through PC
41			Exhibit social signals to HC
42			Move to CPS Layer

2.3. Resource Optimization

We introduced a decision variable, "X", for each possible assignment of resources to the situations. In general, we can say that any decision variable X_{rs} equals "1" if resource $r \in R$ is assigned to the situation $s \in S$ or "0" otherwise.

2.3.1. Situation Constraints

We discussed the constraints associated with situations. These constraints ensure that each situation is handled by exactly one resource. This corresponds to the following:

$$r \in R \sum X_{rs} \leq 1 \tag{3}$$

Less than <1 is included to incorporate the null when no resource is assigned to the situation in an iteration.

2.3.2. Resource Constraints

The resource constraint ensures that, at most, one situation is assigned to each resource. However, it is possible sometimes that not all the resources are assigned. For example, if CPS encounters two situations and only one resource is suitable to handle both. Then the situations are handled sequentially by the resource in order of anxiety. We can write this constraint as follows:

$$s \in S \sum X_{rs} \leq 1 \tag{4}$$

This constraint is less than <1 to allow the possibility that a resource is not assigned to any situation.

2.3.3. Objective Function

The objective is to maximize the total matching score (anxiety factor) of the assignments that satisfy both the situation and resource constraints. The objective function can be concisely written as follows:

$$Maximize\ s \in S \sum r \in R \sum a_{rs}\, X_{rs} \tag{5}$$

3. Experimental Validation

Examples of automotive parts assembly and beverage packaging were considered, as shown in Figure 3. The scenario involves a cobot and a human operator performing tasks dependent on information received from multiple sensors integrated with a collaborative CPPS. Different items arrive at workstations. They are then assembled or packaged in a sequence through a cobot. A human supervisor assists in the completion of the process by performing several tasks. In addition to the dedicated tasks, the supervisor also monitors operations for anomalies and erroneous activities, i.e., they are in both collaborative and supervisory roles. As the assembly/packaging process completes, the supervisor gives a command for the next one, and the cobot moves accordingly. The cobot performs specific operations in collaboration with the operator to complete the task, e.g., in the case of packaging, the robot picks bottles and cans in a sequence from specific locations and drops them in specific slots in the crate, whereas in the case of assembly, the robot picks two types of gears from specific locations in a sequence and place them on a case for assembly. The operator, in the case of packaging, replaces the crate on completion of packaging, and in the case of assembly, tightens the top plate with screws and presses a button for the conveyor to deliver the next case. The human supervisor is also responsible for corrective actions on wrong item arrival, wrong sequence, or absence of item from the location. We call this whole process a standard procedure.

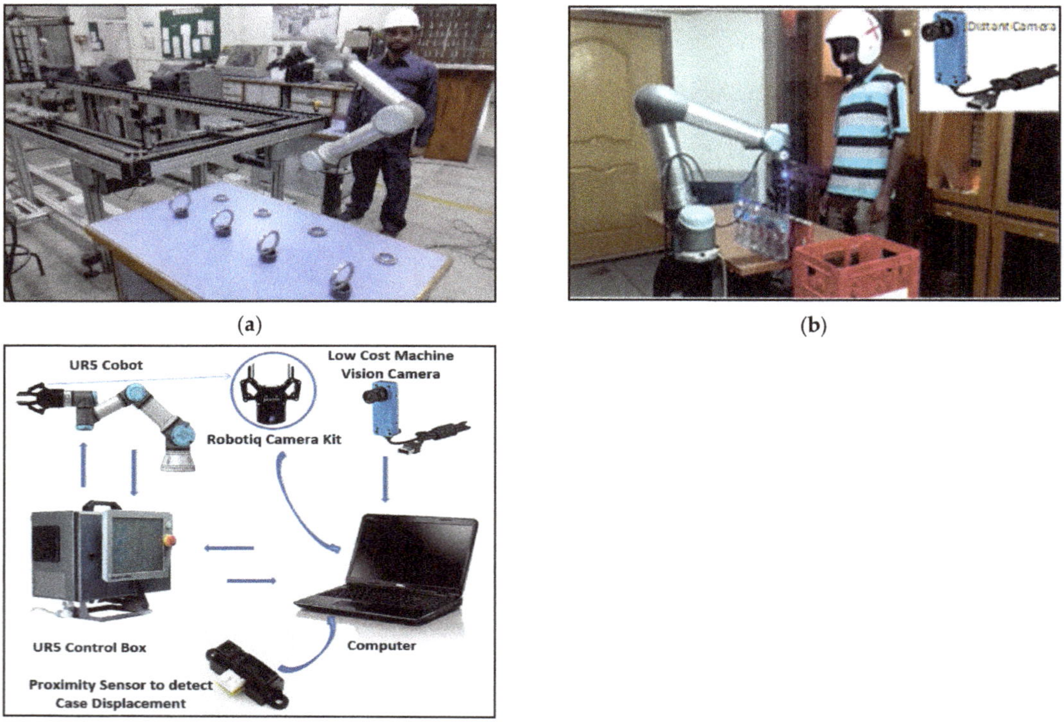

(c)

Figure 3. (**a**) Experimental Case 1; (**b**) Experimental Case 2; (**c**) Interface of CPPS elements.

There may be situations when the process may not proceed as intended, and it may face various unwanted and unforeseen situations. In order to cater to this, different situations were anticipated for both the cases that can emerge during a cycle; these include the intended situation as well as unwanted situations. The considered situations are as follows: the right item is the main task/intended situation i.e., the right items (beverages/gears) are in place for pick and place operations/assembly; the wrong item is the item at the work location is either not in list or wrong in sequence; no item is when no item appears at the work location; the human interference is when operator interferes in any task at any location, as they find that the robot may not be able to perform the task or they find any anomaly; displaced case/crate is that the operation of packaging/assembly cannot be completed when the crate/case is displaced from the designated location; unidentified person is that any unknown person in the workspace is a hazard to the system and to themselves; foreign object is any object not required in the workspace is also a hazard to the system; obsession is any situation actually not affecting but disturbing the outcome of the system, and it is generally established after few iterations when the operator realizes that the situation is a false alarm, e.g., in both the case studies the object detection algorithm detects the table on which the items are placed as a foreign object; time delay is the completion time variance in the intended operation; threshold distance is the breach of minimum distance that is established to be safe for collaboration between the human and the robot; cobot power failure is when the cobot stops work due to power failure; cobot collision is when the cobot collides with the human operator. Different situational awareness techniques were used to detect the listed situations. The object detection technique is used to detect a right item, a wrong item, no item, an unidentified person, and foreign objects. YOLOv3 [48] is trained to detect gears, bottles, and cans along with day-to-day general objects. RFID sensor in the helmet is used to identify the authorized operator, an IR proximity sensor to detect the displaced case/crate, impact sensors in the cobot to detect cobot collision, a power sensor to detect cobot power failure, and the clock to gauge the time delay. The human interventions are detected through a pose estimation algorithm which takes the feed through a camera installed above the workspace. Open pose [49] is used to detect different human poses. The separation distance between the cobot and the operator is used to evaluate the threshold distance; this was performed by calculating the distance between the center of the cobot and the operator detection bounding box. The question is how to map these situations to calculate the anxiety index and anxiety factor and then analyze and decide on requisite actions on them. As soon as any of the listed situations are detected through the visual sensing/other detection techniques, an input of detection is given to the main program. The situation's severity, anxiety index, variables relative to resources, and anxiety factor were already evaluated before the commencement of the operation are taken into account. The severity I for each situation was assessed by the experts by assigning weights to each situation against all other situations in the Ishikawa diagram, as shown in Figure 4. "1" was assigned to the other situation, which was decided to have low priority than the main situation and "0" if it had high priority. The weight was assigned based on the voting of experts. The ranking of the situation (index I) was the total of weights assigned under it. The anxiety A was then calculated for each situation by putting in values of categories and severity as stated in Table 1. There were two resources, a human and a cobot; the matching score (a_{rs}) was then estimated as explained in Section 2 for both the resources vs. the situations and fed into the central database. The severity, categories, the value of anxiety evaluated for the cases, and the anxiety factors are shown in Table 3.

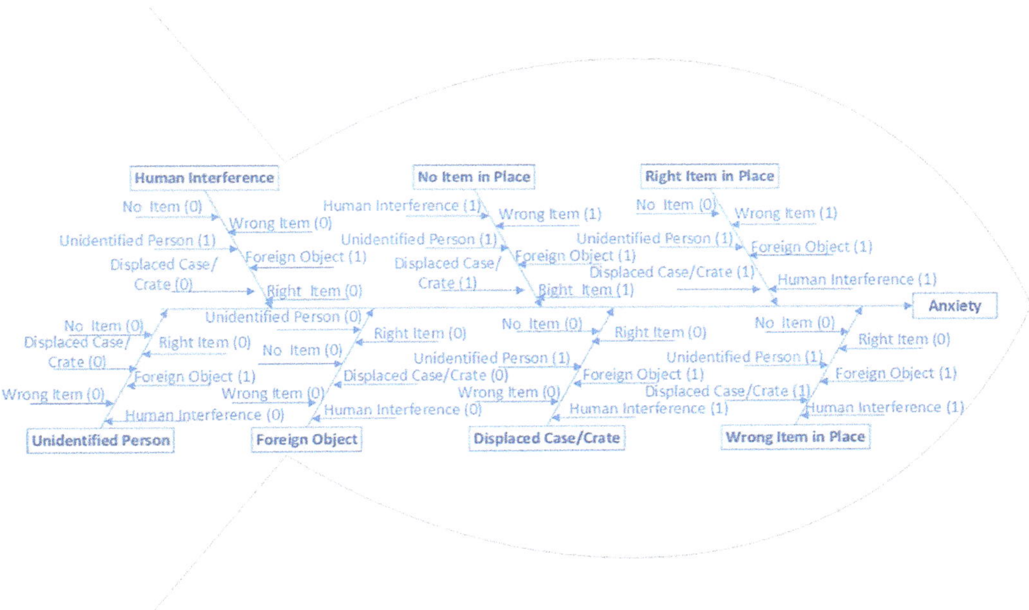

Figure 4. Calculation of index I for situations through Ishikawa.

Table 3. Expected situations, their anxiety, variables, and anxiety factors.

Situation	Severity (*I*)	Category	Parameter (*A*)	Task Variable (*t*)		Preference Variable (*p*)		Resource Suitability Variable (*Q*)		Anxiety Factor (*a*)	
				Human	Cobot	Human	Cobot	Human	Cobot	Human	Cobot
Right item	27	*G*	20	1	0	4	4	1	1	23	24
Wrong item	45	*K*	38	1	0	6	6	1	1	43	44
No item	36	*K*	34.4	0	1	5	5	1	0	39.4	0
Human interference	63	*K*	45.2	1	0	8	8	0	1	0	53.2
Displaced case/crate	54	*K*	41.6	1	0	7	7	1	1	47.6	48.6
Unidentified person	72	*K*	48.8	0	1	9	9	1	0	57.8	0
Foreign object	81	*K*	52.4	0	1	10	10	1	0	62.4	0
Obsession	0	*O*	0	1	1	1	1	0	0	0	0
Time delay	9	*N*	1.8	0	1	2	2	1	0	3.8	0
Threshold distance	18	*N*	3.6	0	1	3	3	1	0	6.6	0
Cobot Power Failure	100	*P*	100	0	1	12	12	1	0	112	0
Cobot Collision	91	*P*	98.2	0	1	11	11	1	0	109.2	0

The main program was divided into five modules to implement the approach on the case studies, which are the main module, item in place module, anxiety module, optimization module, and the decision-making module. The main module holds the other modules. This module keeps count of the operations in the cycle and performs the intended task. It looks for the situations during each iteration and ascertains the matching score through the anxiety module in case multiple situations emerge. The item-in-place module is specific to each particular case and checks whether the right or wrong item is in place through object detection and item count. A machine vision camera was placed to detect the objects. The module adapts the contingency plan if the right item is not in place through the decision-making module. The optimization module assigns the resources to the situations by identifying the highest matching score through MIP using Equations (3)–(5). By analyzing Table 3, we can see that the first basis for the resource assignment is the resource suitability variable Q; if it is zero, the resource cannot be assigned to the situation. The second basis is the task variable t; in our case, 0 was assigned to the preferred resource. Hence the resource with $t = 0$ was assigned if it was not already committed and if the resource suitability variable was not "0". The third basis is the preference variable p, the situation with the higher value was addressed first by the two resources, and the remaining was addressed subsequently after the disposal of the initial ones. The cumulative effect of all these variables is the anxiety factor. The table gives a clear depiction of when any situation appears, which out of two resources can be assigned, which the prior resource is, and which situation is addressed first. The decision-making module then decides the contingencies for the identified situations through actions performed by the resources. The actions are either performed by the cobot or the human supervisor in our scenario, which is dependent on the specific role assigned to the resource for a particular case. The roles assigned to the situations are: for the right item, the robot picks up and places the item at designated location, the human can do so but the preference is given to the robot; for the wrong item, the robot picks up the item and places it at the spot dedicated for them, the count, however, is not increased and the human has to place the right item at the location; in the case of no item in location, the robot moves to the next location and the human places the item at the drop point; if the human operator interferes due to any anomaly in the process or defected item, they may drop the item at the drop point or the wrong item spot, the count is then incorporated by checking the human pose when performing the action; for a displaced case/crate, the robot adjusts it by pushing it to the fixed enclosure, and if the robot is not available, then the human performs it; if an unidentified person enters the workspace, the human has to remove them from the area, and a caution is displayed on the screen for the human to perform this action; similarly, if a foreign object appears in the workspace, a caution is raised to the human to remove the object from the workspace, and in both the last two cases, the robot stops action until the human presses the button for resume operation; in the case of obsession, none of the resources perform any action and perform the task as intended; if the time delay and threshold distance is breached, a caution is given to the human operator to analyze and adjust accordingly; in case of cobot power failure, the human checks the reason, rectifies it, and resumes the operation, however, the whole system is to remain at stand still; similarly in case of cobot collision, the whole system comes to stop and the human checks, resolve the issue, and then resumes the operation. When analyzing, we see that all the actions are interlinked with the variable assigned in Table 3. The experts actually decide the values of the variables based on experience and consultation.

3.1. Survey

We carried out a survey that consisted of participants pooled from university students and faculty. The ages of the participants range from 19 to 40 years (M = 28.7, SD = 7.73).

The participants were informed before the experiment to monitor the screen for instructions during the operation continuously. We devised our own metrics similar to the work in [43]. The subjects had nominal knowledge of the specifics of the system and were briefed on the assigned task only. A total of 20 subjects participated, the repeated-measure design was considered for the experiment, and subjects were divided into two groups randomly. Two types of conditions were assigned, the first group included those who first worked with the decision-making design (n = 11), and the second group included those who worked with the standard design without a decision-making system (n = 9). The subjects were not informed before the experiment what condition was assigned and what metrics were being measured. We measured both quantitative and subjective measures. The quantitative measures include the decision time taken to decide on each situation and the accuracy of the process. The subjective measures include perceived safety, comfort, and legibility based on the questionnaire responses.

First, both groups executed a training round, during which the participants performed the complete experiment by themselves, without an assistant and the robot, to familiarize themselves with the task. Next, all participants were provided with a human assistant to perform the task collaboratively as would be required to perform with a robot in the subsequent phase, i.e., the 3rd phase. Finally, two task executions were conducted firstly in one condition and later in a switched mode; however, the sequence was different for both groups. A questionnaire was given to participants after each task execution. Each participant performed two training sessions with the robot before each task execution to build mental compatibility. In order to prevent any involuntary bias from the participants, the first task execution was conducted in a way that the participant was unaware of which out of two conditions each participant had been assigned to. Before the conduct of the alternate mode, participants were educated that the system would behave differently during the second phase. The participants were briefed; the robot could take some automated measures, and they were required to monitor instructions on the screen.

The questions, shown in Table 4, were intended to determine each participant's satisfaction with the robot as a teammate as well as their perceived safety, comfort, and legibility. A 5-point Likert scale was used for the two questionnaires on which the participants had to respond, strongly disagree to strongly agree for the first questionnaire and much less to much more for the second questionnaire. Based on the dependent measures, the two main hypotheses in this experiment were as follows:

Table 4. Questionnaires.

	Fluency with the Collaborative Robot
1	I trusted the robot to do the right thing at the right time.
2	The robot did not understand how I desired the task to be executed.
3	The robot kept disturbing me during the task.
4	The robot and I worked together for better task performance.
	Perceived safety, comfort, and legibility
5	I felt safe while working with the robot.
6	I trusted the robot would not harm me.
7	The robot moved too drastically for my comfort.
8	The robot endangered the safety of unknown persons in the workspace.
9	I understand what the robot will be doing ahead.

Hypothesis 1. *Using a decision-making framework for anxiety will lead to more fluent human–robot collaboration based on timely automated decisions and the accuracy of the approach.*

Hypothesis 2. *Participants will be more satisfied with the decision-making framework performance while collaborating with the robot and will feel more comfortable, safe, and legible compared with a CPPS that uses standard task planning.*

The automated decision time cannot be calculated for both conditions as the standard approach does not cater to unprecedented situations; the operator has to stop the system and apply countermeasures. The accuracy of the approach is defined as the number of errors (situations that could not be handled automatically) observed during one cycle divided by the total number of iterations. At least 2 situations were intentionally generated in each trial to check the system response.

3.2. Results

The detection of the objects through the object detection technique is shown in Figure 5.

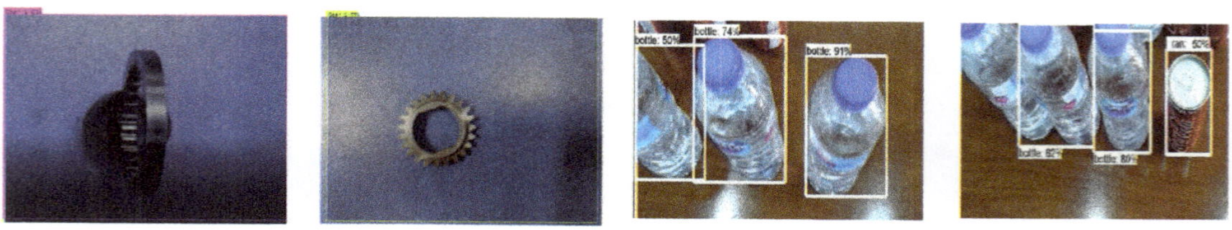

Figure 5. Detection of objects through object detection algorithm, (**a**) gear 1; (**b**) gear 2; (**c**) bottles; (**d**) cans.

Similarly, human detection was also carried out through the object detection technique. The results for the detection of an authorized operator and an unidentified person are shown in Figure 6. The identification of the authorized operator within the workspace through object detection was complemented by the RFID sensor in the helmet.

(a) (b) (c)

Figure 6. Detection of persons through object detection algorithm, (**a**) operator in blue and unidentified person in red; (**b**) authorized operator; (**c**) operator and unidentified person.

Detection of specific poses of the operator is shown in Figure 7, the initial is the detection of the pose of the operator when interfering in assembly, and the latter two are the poses of the operator while placing objects himself. It is pertinent to mention here that the contingencies were adopted only when a particular pose detected is complemented by the position of the robot at the same location. This is because, while performing parallel operations, the same pose could be detected while the cobot may be operating at some other place.

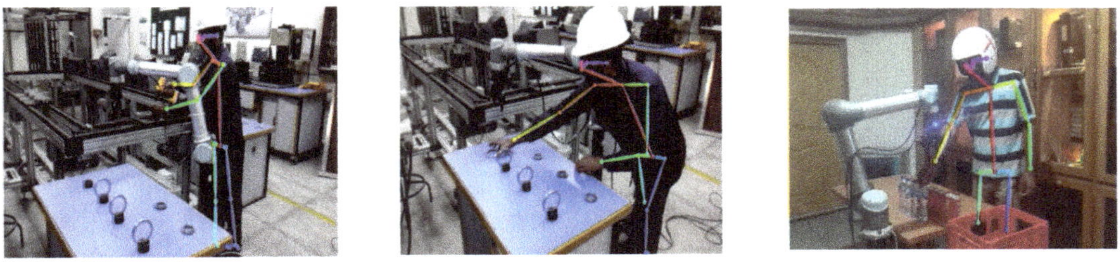

(a) (b) (c)

Figure 7. Detection of specific poses through pose detection algorithm, (**a**) pose while interference in assembly; (**b**) pose while interference in picking of gears; (**c**) pose while interference in picking of beverages.

Our approach is capable of detecting multiple situations and their disposal at once. As an example, two individual situations were considered at a time, which are a displaced case/crate and an unidentified person. An inspector entered the workspace and displaced the outer case in the first scenario, and the crate entered the workspace in the second during inspection (see Figure 8).

The time taken for decision for handling situations for two of the cycles in one test run is shown in Figure 9. Anxieties, total anxiety, and the maximum anxiety situation for each iteration are shown. The maximum time taken to decide on a single situation was noted as 0.03 s.

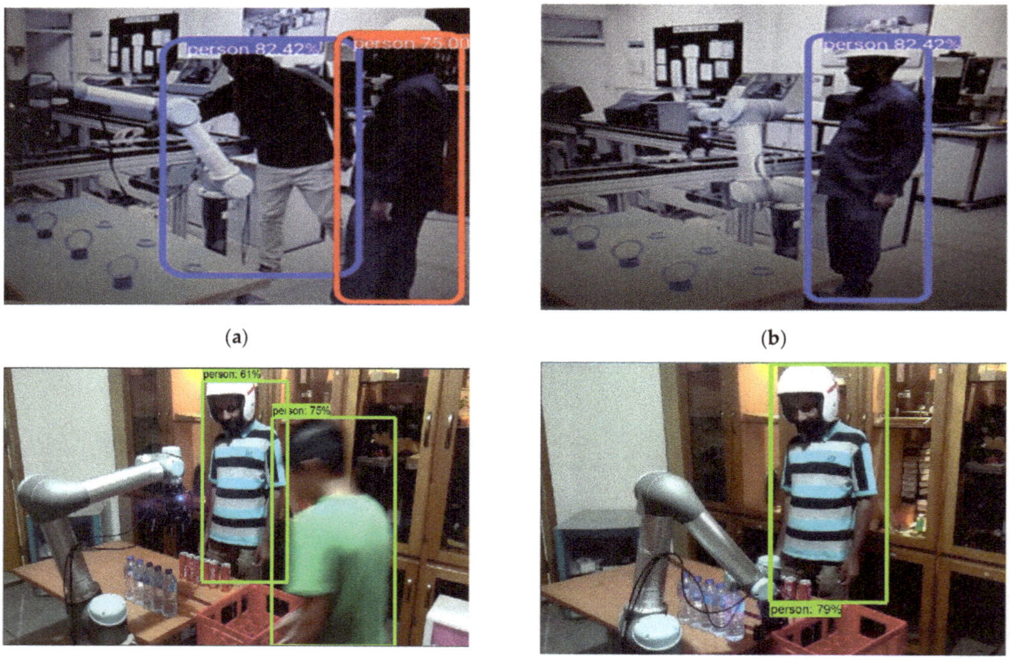

(a) (b) (c) (d)

Figure 8. (**a**) Detection of unidentified person and displaced case; (**b**) Case being aligned by the cobot; (**c**) Detection of unidentified person and displaced crate; (**d**) Crate being aligned by the cobot.

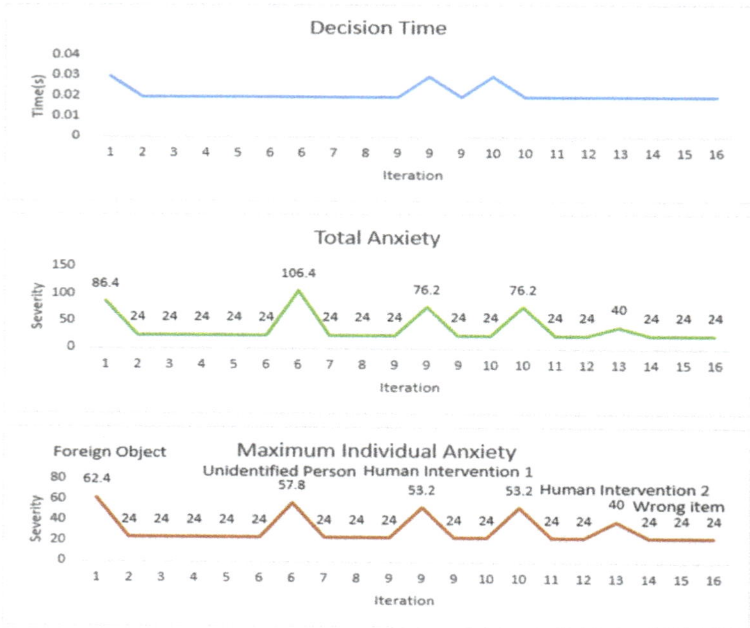

Figure 9. Anxiety evaluation and counter strategy decision time.

The quantitative analysis from the survey was carried out; the overall mean automated decision-time (MADT) for the proposed method and accuracy of both approaches were calculated with a minimum of two situations in each cycle. It was revealed that the mean automated decision time for a contingency is 0.21 s, as can be seen in Table 5. The decision-making system (accuracy of 89.98%) was found to be 16.85% more accurate than the standard system (accuracy of 73.125%). The *t*-test for the significance of the results was conducted with a confidence level set to $p < 0.05$ (95% confidence level), and the *p*-value for the test was found to be $p < 0.01$. The standard error mean (SEM) for the MADT was found to be 9.36×10^{-5}, the upper limit was found to be 0.0211 s, and the lower limit was 0.0208 s.

Table 5. Accuracy of two methods: mean automated decision-time and the SEM.

Subject	Standard Method			Decision-Making Method				Remarks
	Iterations	Errors	Accuracy	Iterations	Errors	Accuracy	MADT	
1	8	2	75	16	1	93.75	0.0206	Average Accuracy (Standard) = 73.12% SEM = 1.02 Average Accuracy (Decision-Making Method) = 89.98% SEM = 0.94 MADT=0.021 s SEM = 9.36×10^{-5} $p = 1.83 \times 10^{-10}$
2	8	2	75	15	1	93.33	0.0206	
3	8	2	75	14	1	92.85	0.0207	
4	8	2	75	17	2	88.23	0.0211	
5	8	2	75	16	1	93.75	0.0206	
6	8	2	75	13	2	84.61	0.0215	
7	8	2	75	12	1	91.66	0.0208	
8	8	3	62.5	14	2	85.71	0.0214	
9	8	2	75	10	2	80	0.022	
10	8	3	62.5	11	1	90.90	0.0209	
11	8	2	75	12	1	91.66	0.0208	
12	8	2	75	10	1	90	0.021	
13	8	2	75	15	3	80	0.022	
14	8	3	62.5	14	1	92.85	0.0207	
15	8	2	75	11	1	90.90	0.0209	
16	8	2	75	12	1	91.66	0.0208	
17	8	2	75	12	1	91.66	0.0208	
18	8	2	75	11	1	90.90	0.0209	
19	8	2	75	13	1	92.30	0.0207	
20	8	2	75	14	1	92.85	0.0207	

Significant differences (at $p < 0.05$) were found for the questions of fluency with the collaborative robot; the participants exposed to the decision-making system agreed more strongly with "I trusted the robot to do the right thing at right time" ($p < 0.01$) and "The robot and I worked together for better task performance" ($p < 0.01$), and disagreed more strongly with "The robot did not understand how I desired the task to be executed" ($p < 0.01$) and "The robot kept disturbing during the task" ($p < 0.01$).

Similarly (at confidence level $p < 0.05$), significant differences were found for the questions of perceived safety, comfort, and legibility; the participants exposed to the decision-making system agreed much more with "I felt safe while working with the robot" ($p < 0.01$), "I trusted the robot would not harm me" ($p < 0.01$), and "I understand what robot will be doing ahead" ($p < 0.01$), and agreed much less with "The robot moved too drastic for my comfort" ($p < 0.01$) and "The robot endangered safety of unknown persons in workspace" ($p < 0.01$).

The results support both the hypothesis in favor of the decision-making system; they indicate that the proposed approach leads to more fluent HRC (Hypothesis 1) and also highlight that the method extends safety, comfort, and legibility (Hypothesis 2).

3.3. Discussion

The research shows that the previous works address limited situations and solely focus on a single aspect, e.g., collision avoidance, motion planning, or psychological safety. The proposed method is seen to collectively address all the issues earmarked by the experts that a CPPS may face. Here, we also wanted to compare the current method with the previous work [46]. Only the most prior situation with maximum anxiety can be handled at once using the previous method, whereas the current approach is optimized to employ all available resources to relieve the current state of anxiety. The method is feasible to include any number of resources by modifying the equations in the optimization algorithm. At times when limited resources are available, situations with higher anxiety were addressed first, followed by lower ones. The current method combines human knowledge and intelligence with AI techniques to minimize the decision time; the maximum time recorded to decide on a scenario is 0.03 s, which shows that the method is not time-intensive. The indexing of anxiety was carried out through a more generic approach in the previous method that lacked rationale for scaling different levels. The current method partially uses the previous technique; however, a logical approach with a biomimetic connection was presented, which makes it easy to differentiate and prioritize situations.

The contribution of the works is that the CPPS was made more intelligent, safe, resilient, and smart. The decision-making of the production system was improved, which provides it the flexibility to tackle multiple situations at once in an optimal manner and can fit in any industrial scenario; manufacturing, assembly, packaging, etc. The amalgamation of four layers highlights the contribution of AI in the industry that increases the overall productivity of the system. The work is a real manifestation of the integration of the computers in Industry 4.0, which is implemented in every layer of the proposed work, i.e., the functioning of the CPS, situational assessment, optimization, and decision-making through software, algorithms, AI, and interface.

The technique in the future can be reinforced by machine learning techniques to make it more intelligent. The CPPS may learn in real-time from the environment and will improve the knowledge base continuously. As the existing system is dependent on predefined solutions, it is proposed that the system may learn through repetitive patterns and solutions provided to the previously confronted situations and may automatically resort to the same solutions. Cloud base and semantic systems are emerging trends that may be incorporated further to enhance the capability and customization of these systems. In this way, multiple situations can be updated and omitted automatically in the system, though supervision by the human component is still recommended. We used machine-learning-based image processing techniques as a tool, which are known for uncertainties due to the statistical methods and probability distributions used in them. The probabilistic nature of these models is prone to errors that cannot be ignored. In the future, the confidence level of detection may be included in the calculation of the anxiety factor. The accuracy of the method may either be increased by selecting a threshold level of detection or adding a variable incorporating the detection level that will affect the anxiety factor of the confronted situation.

4. Conclusions

A framework for social safety was proposed that caters to the physical and cognitive issues of a collaborative robotic system. The framework intelligently safeguards the interest of both human and machine teams with the introduction of the unique concept of the "anxiety factor of CPS". A connective framework was laid out that incorporates anxiety generated by dynamic situations and decides, keeping in consideration previous knowledge, human intelligence, and AI. The method also employs optimization supported by visual and IR sensing techniques that accommodates any number of resources. A flexible system for a collaborative system was thereby produced in which the cooperating elements can respond well to untoward situations. Experimental validation of the method on real-world problems proved its efficacy and applicability. Thus the decision-making of the Collaborative robotic CPPS is improved, which is flexible to cater to multiple situations optimally and can fit in any industrial scenario such as manufacturing, assembly, packaging, etc., ultimately enhancing task efficiency, productivity, and safety.

Author Contributions: Conceptualization, S.O.B.I.; methodology, S.O.B.I.; software, S.O.B.I.; validation, S.O.B.I.; formal analysis, S.O.B.I.; investigation, S.O.B.I.; resources, S.O.B.I.; data curation, S.O.B.I.; writing—S.O.B.I.; writing—review and editing, W.A.L.; visualization, S.O.B.I.; supervision, W.A.L.; project administration, S.O.B.I. and W.A.L. All authors have read and agreed to the published version of the manuscript.

Funding: This research received no external funding.

Conflicts of Interest: The authors declare no conflict of interest.

References

1. Thrun, S. Toward a framework for human-robot interaction. *Hum. Comput. Interact.* **2004**, *19*, 9–24.
2. Bien, Z.Z.; Lee, H. Effective learning system techniques for human–robot interaction in service environment. *Knowl. Based Syst.* **2007**, *20*, 439–456. Available online: https://www.sciencedirect.com/science/article/pii/S0950705107000135 (accessed on 11 September 2022). [CrossRef]
3. Severinson-Eklundh, K.; Green, A.; Hüttenrauch, H. Social and collaborative aspects of interaction with a service robot. *Robot. Auton. Syst.* **2003**, *42*, 223–234. Available online: https://www.sciencedirect.com/science/article/pii/S0921889002003779 (accessed on 11 September 2022). [CrossRef]
4. Murphy, R.R. Human-Robot Interaction in Rescue Robotics. *IEEE Trans. Syst. Man Cybern. Part C (Appl. Rev.)* **2004**, *34*, 138–153. [CrossRef]
5. Nourbakhsh, I.R.; Sycara, K.; Koes, M.; Yong, M.; Lewis, M.; Burion, S. Human-Robot Teaming for Search and Rescue. Available online: https://ieeexplore.ieee.org/abstract/document/1401846/ (accessed on 11 September 2022).
6. Kwakkel, G.; Kollen, B. Effects of robot-assisted therapy on upper limb recovery after stroke: A systematic review. *Neurorehabil Neural Repair* **2008**, *22*, 111–121. [CrossRef]
7. Scassellati, B.; Admoni, H.; Matarić, M. Robots for use in autism research. *Annu. Rev. Biomed. Eng.* **2012**, *14*, 275–294. [CrossRef]
8. Goodrich, M. Human–robot interaction: A survey. *Found. Trends®Hum.–Comput. Interaction.* **2008**, *1*, 203–275. [CrossRef]
9. Fong, T.; Nourbakhsh, I.; Dautenhahn, K. A Survey of Socially Interactive Robots: Concepts, Design and Applications. 2002. Available online: http://uhra.herts.ac.uk/handle/2299/3821 (accessed on 11 September 2022).
10. Chandrasekaran, B.; Conrad, J.M. Human-robot collaboration: A survey. *Int. J. Hum. Robot.* **2008**, *5*, 47–66.
11. Lemaignan, S.; Warnier, M.; Sisbot, E.A.; Clodic, A.; Alami, R. Artificial cognition for social human–robot interaction: An implementation. *Artif. Intell.* **2017**, *247*, 45–69. [CrossRef]
12. Mead, R. Autonomous human–robot proxemics: Socially aware navigation based on interaction potential. *Auton. Robot.* **2017**, *41*, 1189–1201. [CrossRef]
13. Krüger, J.; Lien, T.K.; Verl, A. Cooperation of human and machines in assembly lines. *CIRP Ann. Manuf. Technol.* **2009**, *58*, 628–646. [CrossRef]
14. Lasota, P. Analyzing the effects of human-aware motion planning on close-proximity human–robot collaboration. *SAGE J.* **2015**, *57*, 21–33. [CrossRef] [PubMed]
15. Morato, C.; Kaipa, K.; Zhao, B.; Gupta, S.K. Toward Safe Human Robot Collaboration by Using Multiple Kinects Based Real-Time Human Tracking. 2014. Available online: https://asmedigitalcollection.asme.org/computingengineering/article-abstract/14/1/011006/370115 (accessed on 2 October 2021).
16. Flacco, F.; Kröger, T.; de Luca, A.; Khatib, O. A depth space approach to human-robot collision avoidance. In Proceedings of the 2012 IEEE International Conference on Robotics and Automation, Saint Paul, MN, USA, 14–18 May 2012; pp. 338–345. [CrossRef]
17. Schiavi, R.; Bicchi, A.; Flacco, F. Integration of Active and Passive Compliance Control for Safe Human-Robot Coexistence. 2009. Available online: https://ieeexplore.ieee.org/abstract/document/5152571/ (accessed on 19 June 2022).
18. Safeea, M. Minimum Distance Calculation Using Laser Scanner and IMUs for Safe Human-Robot Interaction. 2019. Available online: https://www.sciencedirect.com/science/article/pii/S0736584518301200 (accessed on 19 June 2022).
19. Vannoy, J. Real-time adaptive motion planning (RAMP) of mobile manipulators in dynamic environments with unforeseen changes. *IEEE Trans. Robot.* **2008**, *24*, 1199. [CrossRef]
20. Yang, Y.; Brock, O. Elastic Roadmaps: Globally Task-Consistent Motion for Autonomous Mobile Manipulation in Dynamic Environments. 2002. Available online: http://citeseerx.ist.psu.edu/viewdoc/summary?doi=10.1.1.102.3248 (accessed on 24 June 2022).

21. Yang, Y.; Brock, O.; Yang, Y.; Brock, O. Elastic roadmaps—Motion generation for autonomous mobile manipulation. *Auton. Robo* **2010**, *28*, 113–130. [CrossRef]
22. Balan, L. Real-time 3D collision avoidance method for safe human and robot coexistence. In Proceedings of the 2006 IEEE/RS International Conference on Intelligent Robots and Systems, Beijing, China, 9–15 October 2006; pp. 276–282. [CrossRef]
23. Haddadin, S.; Belder, R. Dynamic Motion Planning for Robots in Partially Unknown Environments. 2011. Available online https://www.sciencedirect.com/science/article/pii/S1474667016447051 (accessed on 19 June 2022).
24. Haddadin, S. Real-Time Reactive Motion Generation Based on Variable Attractor Dynamics and Shaped Velocities. 2010. Availabl online: https://ieeexplore.ieee.org/abstract/document/5650246/ (accessed on 19 June 2022).
25. de Luca, A. Sensorless Robot Collision Detection and Hybrid Force/Motion Control. 2005. Available online: https://ieeexplore ieee.org/abstract/document/1570247/ (accessed on 19 June 2022).
26. de Luca, A.; Albu-Schäffer, A.; Haddadin, S.; Hirzinger, G. Collision Detection and Safe Reaction with the DLR-III Lightweigh Manipulator Arm. 2006. Available online: https://ieeexplore.ieee.org/abstract/document/4058607/ (accessed on 19 June 2022)
27. Haddadin, S.; Albu-Schäffer, A.; de Luca, A.; Hirzinger, G. Collision detection and reaction: A contribution to safe physica human-robot interaction. In Proceedings of the 2008 IEEE/RSJ International Conference on Intelligent Robots and Systems, Nice France, 22–26 September 2008; pp. 3356–3363. [CrossRef]
28. Grioli, G.; Bicchi, A.; Schiavi, R.; Grioli, G.; Sen, S.; Bicchi, A. VSA-II: A novel prototype of variable stiffness actuator for safe and performing robots interacting with humans. In Proceedings of the 2008 IEEE International Conference on Robotics and Automation, Pasadena, CA, USA, 19–23 May 2008. [CrossRef]
29. Eiberger, O.; Haddadin, S.; Weis, M.; Albu-Schäffer, A.; Hirzinger, G. On Joint Design with Intrinsic Variable Compliance Derivation of the DLR QA-Joint. Available online: https://ieeexplore.ieee.org/abstract/document/5509662/ (accessed on 19 June 2022).
30. Bicchi, A.; Tonietti, G.; Magazine, A. Fast and "Soft-Arm" Tactics. 2004. Available online: https://ieeexplore.ieee.org/abstract/document/1310939/ (accessed on 19 June 2022).
31. de Luca, A.; Flacco, F.; Bicchi, A.; Centro, R.S. Nonlinear Decoupled Motion-Stiffness Control and Collision Detection/Reaction for the VSA-II Variable Stiffness Device. 2009. Available online: https://ieeexplore.ieee.org/abstract/document/5354809/ (accessed on 19 June 2022).
32. Hirzinger, G.; Albu-Schaffer, A.; Hiihnle, M.; Schaefer, I.; Sporer, N.; Oberpfaffenhofen, D. On a new generation of torque controlled light-weight robots. In Proceedings of the 2001 ICRA IEEE International Conference on Robotics and Automation (Cat No.01CH37164), Seoul, Korea, 21–26 May 2001. [CrossRef]
33. Sharkawy, A.-N.; Koustoumpardis, P.; Aspragathos, N.A.; Koustoumpardis, P.N.; Aspragathos, N. Human–robot collisions detection for safe human–robot interaction using one multi-input–output neural network. *Soft Comput.* **2020**, *24*, 6687–6719 [CrossRef]
34. Rani, P.; Sarkar, N.; Smith, C.A.; Kirby, L.D. Anxiety detecting robotic system—Towards implicit human-robot collaboration *Robotica* **2004**, *22*, 85–95. [CrossRef]
35. Picard, R.W. Affective computing: Challenges. *Int. J. Hum. Comput.-Stud.* **2003**, *59*, 55–64. [CrossRef]
36. Kulić, D.; Croft, E. Pre-collision safety strategies for human-robot interaction. *Auton. Robot.* **2007**, *22*, 149–164. [CrossRef]
37. Bethel, C.; Burke, J. Psychophysiological Experimental Design for Use in Human-Robot Interaction Studies. 2007. Available online: https://ieeexplore.ieee.org/abstract/document/4621744/ (accessed on 19 June 2022).
38. Bethel, C.; Salomon, K. Preliminary results: Humans find emotive non-anthropomorphic robots more calming. In Proceedings of the 4th ACM/IEEE International Conference on Human Robot Interaction, La Jolla, CA, USA, 11–13 March 2009; pp. 291–292 [CrossRef]
39. Chadalavada, R.; Andreasson, H. Bi-Directional Navigation Intent Communication Using Spatial Augmented Reality and Eye-tracking Glasses for Improved Safety in Human–Robot Interaction. 2020. Available online: https://www.sciencedirect.com/science/article/pii/S0736584518303351 (accessed on 19 June 2022).
40. Lasota, P.; Rossano, G. Toward safe close-proximity human-robot interaction with standard industrial robots. In Proceedings of the 2014 IEEE International Conference on Automation Science and Engineering (CASE), New Taipei, Taiwan, 18–22 August 2014 pp. 339–344. [CrossRef]
41. ADragan, D.; Bauman, S.; Forlizzi, J.; Srinivasa, S.S. Effects of Robot Motion on Human-Robot Collaborationi. In Proceedings of the 2015 10th ACM/IEEE International Conference on Human-Robot Interaction (HRI), Portland, OR, USA, 2–5 March 2015; pp. 51–58. [CrossRef]
42. Rad, C.; Hancu, O.; Takacs, I. Smart Monitoring of Potato Crop: A Cyber-Physical System Architecture Model in the Field of Precision Agriculture. 2021. Available online: https://www.sciencedirect.com/science/article/pii/S2210784315001746 (accessed on 1 October 2021).
43. Cyber-Physical Production Systems: Roots, Expectations and R&D Challenges. 2014. Available online: https://www.sciencedirect com/science/article/pii/S2212827114003497 (accessed on 1 October 2021).
44. SGreen, A.; Billinghurst, M.; Chen, X.; Chase, J.G. Human-Robot Collaboration: A Literature Review and Augmented Reality Approach in Design. *Int. J. Adv. Robot. Syst.* **2008**, *5*, 1–18. [CrossRef]
45. D'Auria, D.; Persia, F. A collaborative robotic cyber physical system for surgery applications. In Proceedings of the 2017 IEEE International Conference on Information Reuse and Integration (IRI), San Diego, CA, USA, 4–6 August 2017; pp. 79–83. [CrossRef]

46. Islam, S.O.B.; Lughmani, W.A.; Qureshi, W.S.; Khalid, A.; Mariscal, M.A.; Garcia-Herrero, S. Exploiting visual cues for safe and flexible cyber-physical production systems. *Adv. Mech. Eng.* **2019**, *11*, 12. [CrossRef]
47. Gurobi Optimization Inc. 2021. Available online: https://www.gurobi.com/ (accessed on 1 October 2021).
48. Redmon, J.; Farhadi, A. YOLOv3: An Incremental Improvement. 2018. Available online: http://arxiv.org/abs/1804.02767 (accessed on 1 October 2021).
49. Cao, Z.; Hidalgo, G.; Simon, T. OpenPose: Realtime Multi-Person 2D Pose Estimation Using Part Affinity Fields. 2019. Available online: https://ieeexplore.ieee.org/abstract/document/8765346/ (accessed on 2 October 2021).

machines

MDPI

Article

Curriculum Design and Sim2Real Transfer for Reinforcement Learning in Robotic Dual-Arm Assembly †

Konstantin Wrede [1,*], Sebastian Zarnack [1], Robert Lange [1], Oliver Donath [1], Tommy Wohlfahrt [1] and Ute Feldmann [2]

1 Fraunhofer Institute for Integrated Circuits IIS, Division Engineering of Adaptive Systems EAS, Muenchner Strasse 16, 01187 Dresden, Germany
2 Institute of Control Theory, TU Dresden, Georg-Schumann-Straße 7a, 01187 Dresden, Germany
* Correspondence: konstantin.wrede@eas.iis.fraunhofer.de

† This paper is an extended version of our paper published in Proceedings of the 4th IFSA Winter Conference on Automation, Robotics and Communications for Industry 4.0/5.0 (ARCI 2024), 7–9 February 2024, Innsbruck, Austria.

Abstract: Robotic systems are crucial in modern manufacturing. Complex assembly tasks require the collaboration of multiple robots. Their orchestration is challenging due to tight tolerances and precision requirements. In this work, we set up two Franka Panda robots performing a peg-in-hole insertion task of 1 mm clearance. We structure the control system hierarchically, planning the robots' feedback-based trajectories with a central policy trained with reinforcement learning. These trajectories are executed by a low-level impedance controller on each robot. To enhance training convergence, we use reverse curriculum learning, novel for such a two-armed control task, iteratively structured with a minimum requirements and fine-tuning phase. We incorporate domain randomization, varying initial joint configurations of the task for generalization of the applicability. After training, we test the system in a simulation, discovering the impact of curriculum parameters on the emerging process time and its variance. Finally, we transfer the trained model to the real-world, resulting in a small decrease in task duration. Comparing our approach to classical path planning and control shows a decrease in process time, but higher robustness towards calibration errors.

Keywords: reinforcement learning; curriculum learning; simulation-to-reality transfer; dual-arm robots; robotic assembly; peg-in-hole assembly

1. Introduction

In this paper, we build upon the foundational work presented in our earlier publication "Curriculum-organized Reinforcement Learning for Robotic Dual-Arm Assembly" at the Automation, Robotics & Communications for Industry 4.0/5.0 (ARCI' 2024): 4th IFSA Winter Conference [1]. We aim to expand the original study's scope by performing a simulation-to-reality (Sim2Real) transfer for the proposed dual-arm robotic assembly task. This extension not only broadens the theoretical framework but also incorporates additional experimental results, enhancing the relevance and applicability of our findings.

In modern manufacturing, assembly consumes about 50% of total production time and accounts for 30% of costs [2]. Especially insertion tasks—critical in the final integration of assemblies in consumer electronics like mobile phones—account for 60% of manual human labor time [3]. Assembly tasks typically involve multiple sequential insertion operations, making the use of dual-arm robots advantageous as they can eliminate the need for reorientation and reclamping of workpieces, thus reducing overall process time. However, this solution adds layers of complexity to the control problem. Traditional robotic systems for such assembly are hindered by high costs associated with image processing systems and force/torque measurement sensors, as well as the involvement of human programming. They are also limited regarding the adaptability to varying workpiece geometries, initial inaccuracies, and tight clearances.

In response, this work introduces a strategy utilizing reinforcement learning (RL) to train control algorithms for such tasks. Our approach not only aims at predicting the average process time but also models the variability and resulting distribution introduced by uncertainties in simulation.

We set up two Franka Panda robots to collaboratively perform a square peg-in-hole task. Training control systems in reality can be slow due to its confinement to real-time operations, costly because it necessitates multiple hardware units for parallelization, and risky due to potential damage during exploration phases. Consequently, simulations have become an essential element in the control design stage. We use the Python library robosuite [4], which interfaces with the MuJoCo physics engine [5]. Our control architecture adopts a hierarchical approach, with a low-level impedance control receiving desired trajectories from a central policy trained with RL. The RL training methods are implemented via Stable Baselines3 (SB3) [6] and utilize a sparse reward function proposed by robosuite. The training process is designed as a reverse curriculum: we begin the training with the insertion task nearly completed. As the policy's performance improves, we progressively enhance the difficulty of the task until it reaches a configuration that is realistic for practical applications in terms of the initial peg–hole distance. To ensure the policy's adaptability to varying initial robot configurations, we incorporate domain randomization (DR).

Nevertheless, a significant challenge persists, known as the simulation–reality gap, which complicates the transfer of control systems from simulation to real-world applications [7]. This gap often results from discrepancies such as model mismatches (e.g., inaccurate dynamic parameters) and disturbances affecting the inputs and outputs in the actual system. Additionally, RL faces difficulties in terms of interpreting the trained policy's actions and adapting to changing system conditions without considerable retraining. These issues underscore the importance of exploring the robustness of Sim2Real transfer. We assess the efficacy of this policy in simulation compared to traditional control methods and compare its performance in real-world to the simulation environment.

In this study, we address the research question of how to train a dual-arm robotic assembly policy in simulation and transfer it to a real-world setup. How robust performs the resulting control system to varying tasks in reality, especially compared to a classical method? A more detailed insight into the particular task as problem formulation is provided by Section 3. To approach these research questions, we have structured our investigation into the following points:

- We establish a simulation and training environment for dual-arm robotic assembly using robosuite, coupled with the MuJoCo physics engine.
- We integrate the simulation with SB3 RL methods, employing the OpenAI Gym standard [8].
- We design a training methodology based on a reverse curriculum, enhancing both convergence rates and sample efficiency.
- We investigate how variations in curriculum parameters influence the process characteristics observed in simulations.
- We transfer the trained policy to a real-world setup, comparing its performance to simulation and traditional control strategies.

The remainder of this paper is organized as follows. The following Section 2 introduces related work in the fields of simulation environments, robotic assembly control strategies, and curriculum learning as well as Sim2Real transfer. We use Section 3 to concisely specify the task of this study. In Section 4, we detail the methodological foundation of our research, describing the interplay between low-level controllers and the policy, the reverse curriculum design for training and the eventual Sim2Real transfer. Section 5 first examines the setup and outcomes of experiments in simulation. It explores the transfer of the trained policy to a real-world setting and compares results to those achieved using traditional control strategies. Finally, Section 6 discusses implications of these results, before Section 7 sums up our work, highlights limitations, and derives potential directions for future research.

2. Related Work

In this section, we first present previous work on relevant methods and techniques used in our study, before we provide more details on their implementations in the context of our specific assembly task in Section 4.

2.1. Reinforcement Learning

Reinforcement learning (RL) is a branch of Machine Learning that seeks to mimic the learning processes observed in both humans and animals. Unlike traditional Machine Learning approaches that rely on predefined datasets, RL involves the dynamic generation of data through interaction within a specific environment. In this framework, an agent engages with the environment by performing actions, guided by a reward system that informs the suitability of its decisions based on the currently observed system state. This continuous interaction facilitates the development of a policy, which decides on the agent's responses to different environmental conditions. Such policies may be deterministic, prescribing explicit actions, or stochastic, suggesting actions with certain probabilities. The overarching goal of the agent is to optimize its reward accumulation, thereby progressively refining its decision-making process by training and adapting to novel states within the environment. Consequently, the essence of an RL system is encapsulated in the interaction between the agent and its environment [9].

2.2. Robust Sim2Real Transfer

Transferring policies trained in simulation environments with RL to reality is a necessary step towards actual application of learning-based robotic systems. This process, known as Sim2Real transfer, faces the challenge of being robust to deviations, uncertainties, and unforeseen situations, known as Sim2Real gap, between the two domains. We address this challenge at different stages from decisions on the control system design until deployment.

First, such robustness by architecture can be achieved through a hierarchical control model. This structure integrates classical controllers for low-level hardware actuation with a high-level policy framework. The high-level policy is designed to process all measured values as observations, subsequently generating setpoints for the low-level control. By this architecture the action space is adapted for an accelerated exploration during training and improved safety at deployment. Furthermore, the robustness of the Sim2Real transfer with respect to model mismatch is increased due to decoupling of the policy and system dynamics [10]. In this paper, leveraging experience from past research, as outlined in Section 2.4, we specifically select a suitable low-level controller, as explained in Section 4.1.

Furthermore, it is possible to improve the robustness at the training stage in simulation by applying DR. This technique introduces variations in the simulation settings, such as initial conditions, to more accurately reflect the distribution of real-world data, bridging the gap created by biases between simulated and actual environments. Additionally, DR allows for the modeling of time-varying behaviors that result from system drifts or degradation [11]. This is achieved through dynamics randomization by altering the system's dynamic parameters. To prevent a distribution shift between simulation and reality, it is crucial to consider the diversity of randomization within a dataset—not merely the quantity of samples [12]. In our work, we apply DR described in Section 4.4 matching to a task-specific curriculum with its general concept explained in Section 2.5 and task-specific implementation as a reverse curriculum in Section 4.3.

During the deployment stage on real hardware, the robustness of Sim2Real transfer can be further enhanced by employing methods such as policy-level action integration (PLAI). While dynamics randomization is a common strategy to address model inaccuracies and improve robustness, it often requires extensive additional training and may compromise precision eventually. Drawing inspiration from PID control, PLAI modifies incremental setpoints by integrating policy actions based on the previous desired state rather than the current state. This adaptation significantly improves the control system's ability to reject disturbances and minimize steady-state errors, thereby enhancing overall performance [13]. Our particular utilization of PLAI is detailed in Section 4.5.

2.3. Simulation Environment

For robotics simulations, the gym interface often serves as the cornerstone for connecting with RL toolboxes. A notable example is PyBullet [14], a lightweight yet robust simulation engine, which has been successfully integrated into various RL frameworks, including PyTorch [15]. This integration is used in several studies [16–18]. Additionally, NVIDIA's Isaac Sim has gained traction as a robotics simulator within the RL community, distinguished by its comprehensive and so far most advanced GPU support [19]. It commonly interfaces with PyTorch-based libraries such as skrl [20]. Furthermore, past research [21–23] has utilized MuJoCo within the robosuite framework for its efficient low-level control implementations, a lean framework compared to Isaac Sim. Our work leverages this particular setup, capitalizing on its streamlined integration and benefits in development.

2.4. Control of Robotic Assembly

For single-arm peg-in-hole assembly tasks, compliant control systems such as impedance control have been extensively applied, demonstrating their effectiveness across various studies [24–26]. This conventional approach has been successfully extended to dual-arm setups, with notable implementations detailed by [27]. In these configurations, while one robot actively manipulates the peg, the other typically plays a passive role by just holding the hole part with high compliance. However, engaging both robots actively could significantly enhance process time and reliability. This potential improvement has spurred interest in applying RL techniques. RL has been effectively used to generate reference trajectories for established low-level impedance controllers, as shown in recent research [16].

2.5. Curriculum Learning

Curriculum learning is a technique applicable to goal-oriented tasks that do not require prior domain knowledge, depending exclusively on achieving a state indicative of task completion. The method entails a progressive learning process where the robot starts from various initial states, each incrementally increasing in difficulty (easy2hard), often measured by the spatial distance from the goal. Previous research, such as [28], has explored the automatic generation of such learning curricula. In contrast, our approach involves manually defining the learning steps, tailored to our specific needs.

A recent advancement in this field is the introduction of a sampling-based curriculum [13], which involves gradually elevating the lower bound of a task difficulty. The initial system configurations are selected from a distribution determined by this bound. In scenarios involving dual robots, like in our study, setting these difficulty bounds is more complex. The condition for activating DR in the curriculum is crucial, as it is chosen based on the insertion state to prevent the system from initializing into a potential jamming situation. To our knowledge, this work is the first implementation of curriculum-organized RL specifically applied to a dual-arm robotic assembly task. Moreover, the two-phase structure of our curriculum introduces a novel approach to this field.

3. Problem Formulation

Our aim is to develop a control policy in simulation and successfully transfer it to a real-world setup for tactile assembly applications. Specifically we focus on a peg-in-hole problem with squared objects, implying the relative orientation of the parts to be crucial. Figure 1 illustrates our experimental setup in both simulation and reality, featuring squared objects with a 1 mm clearance, and both Frankas' bases positioned 1.4 m apart. The parts used in the experiments were manufactured from PLA material using a 3D printer, with a printing resolution of 0.1 mm—ten times more precise than the clearance. The gripping process of both parts is out of scope of this work and assumed to be performed in advance.

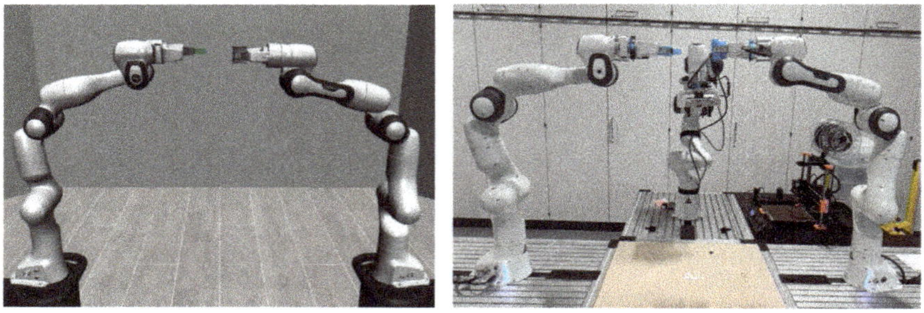

(a) (b)

Figure 1. Environment consisting of two robot arms, peg and hole workpieces. (**a**) Setup in simulation. (**b**) Setup in reality.

4. Methodology

We propose a hierarchically structured control system consisting of two decentral low-level impedance controllers, one per robot, which directly deliver torque commands to the joint motors. A central policy in form of a neural network (NN), trained with RL in a curriculum-based framework, generates setpoint values for these impedance controllers feedbacking current measurements as observations. This architecture is illustrated in Figure 2. To enhance the robustness of the policy and facilitate a successful Sim2Real transfer, DR is applied during specific training sections.

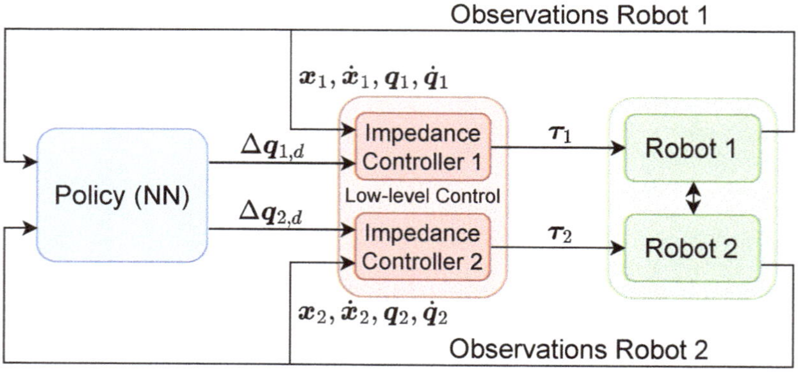

Figure 2. Hierachical control system architecture with a central policy in form of a NN generating setpoints for the decentral low-level impedance controllers.

4.1. *Low-Level Control*

To optimize the action space for the policy network, on each robot, impedance control is implemented, as outlined by [29]. The recommendation to adapt the action space through Cartesian impedance control facilitates rapid policy exploration in simulation. Furthermore, for deploying policies in the real world, the use of joint velocity impedance control is advisable to enhance the efficacy of the Sim2Real transfer. However, to simplify implementation efforts and reduce the need for coordinate transformations, we opted for joint position impedance control. Reducing the amount of Franka Panda specific terms and retaining a mainly robot-agnostic implementation, we omit the moment of inertia matrix, the Coriolis, and centripetal matrix. This leaves a PD-like controller with gravity compensation similar to the joint space position controller with fixed impedance provided by robosuite. To emphasize the compliant behavior of the robot controllers, necessary for small-tolerance insertions, we still refer to this as impedance control. Also, the controller's setpoints are expressed as delta positions. The control law is expressed as:

$$\boldsymbol{\tau} = K_p(\boldsymbol{q}_d - \boldsymbol{q}) - K_d\dot{\boldsymbol{q}} + \boldsymbol{g}(\boldsymbol{q}), \quad (1)$$

$$\boldsymbol{q}_d = \boldsymbol{q} + \Delta\boldsymbol{q}_d. \quad (2)$$

where $\boldsymbol{\tau}$ denotes the joint motor torques, the variable $\Delta\boldsymbol{q}_d$ refers to the desired delta joint positions, and $\boldsymbol{q}$ and $\dot{\boldsymbol{q}}$ denote the measured joint positions and velocities, respectively. Adding the configuration dependend gravity compensation $\boldsymbol{g}(\boldsymbol{q})$ to the control law will be accomplished by the Franka control implementation in the real-world setup internally. The matrix K_p represents the stiffness and K_d the damping as control gains, which are maintained at standard values of the robosuite implementation:

$$K_p = 50\,\mathrm{Nm}\begin{pmatrix}1 & & \\ & \ddots & \\ & & 1\end{pmatrix}, \; K_d = 2\sqrt{50}\,\mathrm{Nms}\begin{pmatrix}1 & & \\ & \ddots & \\ & & 1\end{pmatrix}, \; K_p, K_d \in \mathbb{R}^{7\times 7}. \quad (3)$$

The relationship between these gains is defined by $K_{d,i} = 2\sqrt{K_{p,i}}$, ensuring aperiodical damping and stability.

4.2. *Reinforcement Learning*

We formulate the problem as a Markov decision process, explicitly defining the observation space, action space, and reward function. We utilize a state-of-the-art model-free, off-policy RL algorithm known as Soft Actor-Critic (SAC) [30], employing the implementation provided by SB3. To address the computational demands of the CPU-based physics simulation in MuJoCo and enhance sample efficiency, we integrate a replay buffer using Hindsight Experience Replay (HER) [31]. Table 1 presents relevant hyperparameters involved in the RL training as well as network parameters of the policy.

Table 1. Relevant training and network parameters of the actor-critic policy trained with SAC.

Parameter	Value
Optimizer	Adam
Learning Rate	3×10^{-4}
Discount Factor	0.99
Entropy Coefficient	Auto (init 1.0)
Batch Size	256
Buffer Size	1×10^6
Replay Buffer Size	6
Architecture Actor	1×256 neurons as hidden layer
Architecture Critic	2×256 neurons as hidden layers
Activation Function	ReLU with bias

The observation space $\boldsymbol{O} \ni \boldsymbol{o}$ for our RL setup is based on the default configuration provided by robosuite for the *TwoArmPegInHole* environment:

- Positions and quaternions of the peg and the hole;
- A difference vector and angle φ between the peg and the hole;
- Parallel distance c and perpendicular projected distance d between the peg and the hole;
- The state of each robot, which encompasses the sine and cosine of the joint angles $\boldsymbol{q}$, joint velocities $\dot{\boldsymbol{q}}$, as well as positions and quaternions of the end-effectors (EE).

It is important to note that this observation space does not include any force or torque measurements from additional sensors. The geometric quantities described are illustrated in Figure 3.

Figure 3. Quantities describing the geometric relationship between the peg and the hole.

The action space is defined by the choice of joint position impedance control, specifically targeting the desired deltas of joint coordinates for each robot. The range of these actions is symmetrically scaled by the parameters *output_min* and *output_max*, with the maximum delta values set at:

$$\Delta \boldsymbol{q}_{d,max} = -\Delta \boldsymbol{q}_{d,min} = (0.05,\ 0.05,\ 0.05,\ 0.05,\ 0.05,\ 0.05,\ 0.05)^T \text{ rad}. \tag{4}$$

Limiting the actions in form of delta joint positions given a fixed control period results in a rate limit of the joint velocities. Furthermore, this specification determines the maximum achievable speed for task completion during training and if not scaled during deployment. The numerical limits were empirically established through simulation trials and are subject to further refinement based on the absolute maximum ratings of robot components, work pieces, and the specific requirements of the assembly task.

The reward function utilizes the default sparse reward setup of the *TwoArmPegInHole* environment, structured as follows:

$$r = \begin{cases} 1, & d < d_{max};\ |c| < c_{max};\ \cos\varphi > \cos\varphi_{max} \\ 0, & \text{else.} \end{cases} \tag{5}$$

where $d_{max} = 1$ mm is determined by the part clearance, $c_{max} = 2$ mm, and $\cos\varphi_{max} = 0.95$ rad is set based on intuitive experimentation. This goal-oriented and sparse reward structure is particularly effective for facilitating the application of HER.

4.3. Reverse Curriculum Generation

A reverse curriculum arranges a sequence of related tasks in reverse order, beginning with an initialization close to the goal state and progressing towards more complex initial conditions. This approach leverages the reversible nature of most assembly tasks, aiming to optimize learning efficiency in environments characterized by sparse rewards concentrated in a limited portion of the state space. Under such a curriculum, a policy trained with RL, exploring with randomized actions, is able to quickly accumulate rewards and expedite its learning progression. The policy applies the capabilities acquired at simpler stages to increasingly difficult scenarios. As its competence grows, the initial peg–hole distance or task difficulty is systematically increased to maintain alignment with the policy's improving performance. Specifically, in goal-oriented tasks, the initial system state s_0 can be adjusted to more challenging configurations based on the cumulative reward (return) $R(s_0)$ gathered during inference, thus tailoring the difficulty to the policy's evolving capabilities.

In this study, we structure the curriculum into two distinct phases within each iteration:

1. **Minimum Requirements:** After a predetermined number of training episodes, an inference test determines whether the policy achieves a return R that surpasses a set threshold R_{min}. If successful, the curriculum progresses to the fine-tuning phase. Else, this phase is repeated.
2. **Fine Tuning:** Subsequent to additional training over a fixed number of episodes, another inference is evaluated. If there is a relative improvement R across a specified number of inferences, the fine-tuning phase is extended. If improvement stalls, the difficulty of the initial state is increased, advancing the curriculum to the next iteration, which starts again at the Minimum Requirements phase.

The curriculum is completed successfully by finishing the Fine Tuning step with an initial configuration equivalent to the original task formulation. Figure 4 illustrates this curriculum.

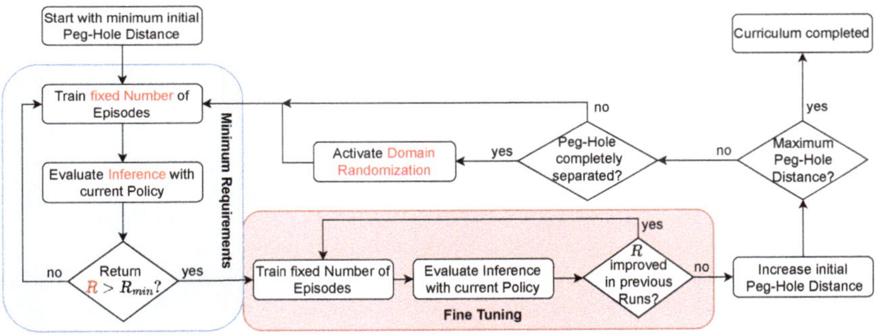

Figure 4. Iterative training procedure of the dual-robot peg-in-hole task as reverse curriculum. Quantities represented by positions marked red will change if DR is activated.

To specify the initial system state, we manually configured the joint angles of both Franka robots as:

$$q_{init} = (0,\ 0.20,\ 0,\ -1.20,\ 0,\ 2.90,\ 0.79)^T \text{ rad.} \tag{6}$$

This setup places the peg tip with $c = 80$ mm away from the hole, defining the baseline for the original task. The curriculum begins with the peg nearly fully inserted into the hole by 24 mm. Initially, during the deterministic phase of the curriculum, which spans 40 iterations and excludes DR, the peg is gradually extracted by 1 mm per iteration at initialization. Following this, starting from a peg–hole distance of 5 mm, DR is applied as white noise to all initial joint angles over the next 15 iterations. During this phase, the peg–hole distance is increased by an average of 3 mm per iteration until the system reaches to the original configuration.

The threshold parameter is set to $R_{min} = 50$ mm, which requires the insertion to be completed for at least a third of the episode, consisting of 150 steps. Choosing the parameter this way strikes a balance between training efficiency and the reliable continuation of the insertion task. In the Fine Tuning phase, the curriculum is designed to allow for three inference evaluations before the reward is considered to have stagnated, indicating proceeding into the next iteration. Each phase of the curriculum involves 100 training episodes in one step. However, an exception is made during the Minimum Requirements phase following the activation of DR. Due to the increased stochastic nature of the environment with DR, the first phase involves not just one but 30 inference evaluations, with returns averaged to account for variability. To ensure an appropriate ratio of training and inference episodes, the number of training episodes per step is increased to 300 during this as long as DR is applied.

4.4. Domain Randomization

In this work, DR plays a critical role in enhancing the policy's ability to adapt to a broad range of initial system configurations. We implement DR by introducing normally distributed white noise with a standard deviation of $\sigma_q = 1.15^\circ$ to all joint angles during initialization. However, to prevent the emergence of undesirable states—such as the peg and hole starting in an condition of both objects penetrating each other—DR is deactivated at first. DR is applied when the initial state involves a complete separation between the peg and the hole. We do not use DR in form of dynamic parameters or sensor measurements, but similar to [13], we choose to implement PLAI instead.

4.5. Deployment on Real-World Hardware

To successfully transfer the trained policy from a simulated environment to the real-world setup, we developed a toolchain that embeds the NN into a control system structured according to Section 4.1. The toolchain for Sim2Real transfer includes the following steps and components:

- **Training the Policy:** The first step involves training the policy within the introduced simulated environment. After finishing the training, the policy is exported as a PyTorch file. This format is chosen for its flexibility and compatibility with various deployment environments.
- **Low-Level Control System:** The low-level joint space impedance controllers are implemented in C++ to interface with the libfranka library on each robot. This ensures real-time control of the robot's joint motors with a frequency of 1 kHz.
- **Connecting both Robots:** Both robots are connected to a central computer that runs the Robot Operating System 2 (ROS 2) [32]. ROS 2 facilitates communication between the robot's specific low-level controllers and the central policy, enabling coordinated actions and reliable low latency data exchange. The motion planning framework *MoveIt 2* [33] is used in tandem with *ros2_control* and RViz to provide an interface for managing and visualizing the robot state as well as an option for classical trajectory planning and execution with interchangeable low-level controllers on the robot hardware. The policy is embedded within ROS 2, providing setpoints for the low-level controllers. This integration allows the high-level policy to guide the robot's actions based on sensory inputs while still meeting real-time constraints. To further streamline the process of integrating a policy into the real-world control system, we developed several ROS 2-based tools. This control utils package provides a simplified python interface for trajectory planning and interfacing MoveIt 2 by wrapping its API utilizing *pymoveit2*.
- **Data Recording:** To evaluate the performance of the policy, we set up a recording and logging mechanisms to capture the joint efforts, joint velocities, and EE forces of every robot arm as well as the total time needed for every episode.

To enhance the robustness of the policy during real-world execution, we apply PLAI to minimize steady-state errors. In simulation, incremental setpoints are based on the current measured joint positions, as stated in Equation (2). In contrast, in the real world setup, we derive these desired joint positions from the previous desired values (last setpoints) and the output of the policy $\Delta \boldsymbol{q}_d(t) = \boldsymbol{\pi}(\boldsymbol{o}(t))$:

$$\boldsymbol{q}_d(t) = \boldsymbol{q}_d(t-1) + \boldsymbol{\pi}(\boldsymbol{o}(t)). \tag{7}$$

This approach enhances the stability and robustness of the policy during real-world execution by reducing the dependency on immediate sensor feedback and instead relying on the consistent application of the policy's outputs over time. The result is a more resilient control strategy that better accommodates the uncertainties and variances inherent in real-world robotic systems (e.g., sensor noise).

5. Results

5.1. Training in Simulation

The primary objectives of the simulation is to train a robust policy for a subsequent Sim2Real transfer and investigate the impact of curriculum parameters on the characteristics and performance of the trained control system. By modifying the distance between both robots and object geometries of robosuite's *TwoArmPegInHole* environment, we establish a simulation consistent with the task description outlined in Section 3.

The moving average of the return R sampled during training is illustrated in Figure 5. The curriculum spans approximately 35,000 episodes to complete. Notably, the reward curve exhibits distinct discontinuities, corresponding to adaptations in the initial joint configuration as the curriculum progresses by iterations. The deterministic learning phase concludes around 17,000 episodes, marked by a significant increase in difficulty due to the activation of DR. This change leads to a slower ascent in the learning curve. With noise introduced at initialization, changes to the joint configuration caused by curriculum progression begin to exert less influence on the task difficulty. After 28,000 episodes, the reward curve plateaus and eventually begins to decline, reflecting the challenges posed by increased initial distances and the speed limitations imposed by actuation constraints. The specified maximum episode length is just sufficient to achieve the threshold $R_{min} = 50$ mm.

Following the training phase, the control policy was evaluated across 10,000 inference episodes using DR at the original initial task configuration. The histogram shown in Figure 6 illustrates the distribution of the process cycle time derived from the simulation step, which posed the first positive sparse reward. On average, the task is accomplished within 4.99 s with a standard deviation of 0.80 s. Notably, the insertion task succeeds in 99.8% of all inference episodes in less than 7.5 s, demonstrating a high reliability and effectiveness of the trained control system in simulation.

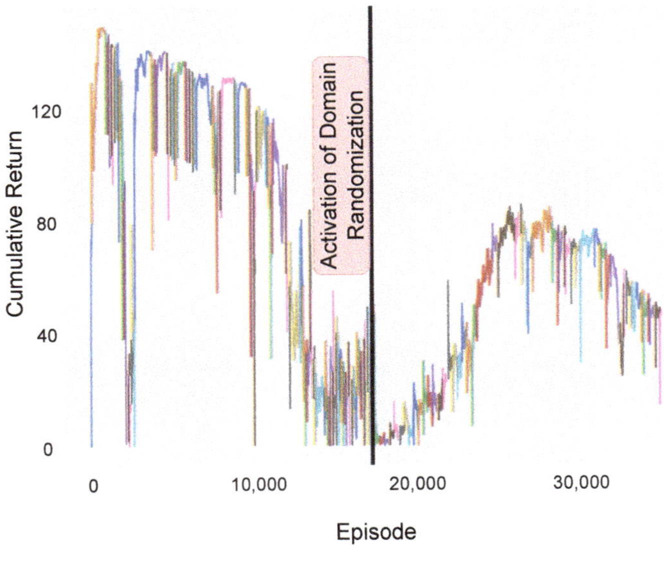

Figure 5. Return per training episode. Each color highlights a new iteration of the curriculum.

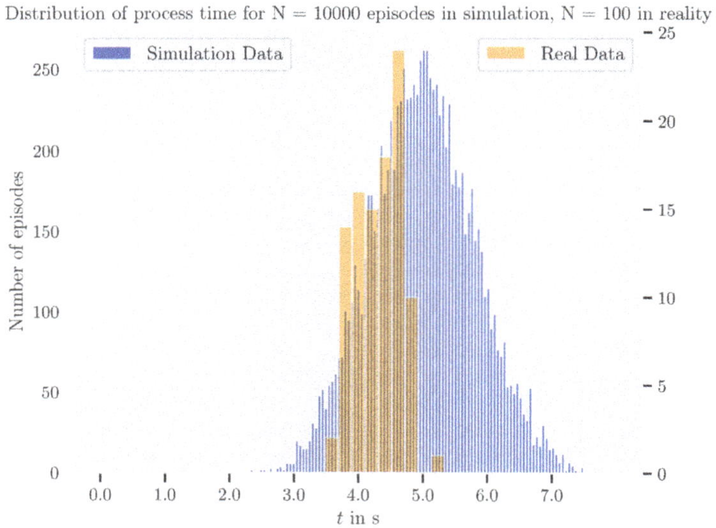

Figure 6. Comparison of process time for the proposed RL-based control system in simulation and reality.

5.2. Transfer to Reality and Evaluation

To evaluate the effectiveness of the trained control policy beyond simulation, we deployed it on the physical robots leveraging the methods explained in Section 4.5. During 100 inference episodes initiated from the original joint configuration stated in Equation (6), the policy achieved a success rate of 100%. Figure 6 compares the process times for both the simulated and real-world setup. In reality, the average process time was measured as 4.31 s, with a standard deviation of 0.34 s, both smaller than in simulation. This could be caused by overestimating uncertainties in the initial configuration due to measurement noise and limited encoder resolution by a high amount of DR in simulation, while having a very narrow distribution of actual initial states close to the nominal original setup in reality. Additionally, 48% of all real-world inferences completed the insertion within this average time. The high success rate of 100% in reality proves the robustness of the trained policy, confirming its effectiveness in real-world settings for the original task configuration.

Additionally, we compared a classical trajectory planning approach using MoveIt 2 to our RL-trained policy. The classical control method, which employs low-level PD joint controllers, used a pre-defined trajectory with parameters noted in Table 2.

Table 2. Planning parameters for a classical joint trajectory controller.

Parameter	Value
Planning Time	1 s
Planner	RRTConnect
Max. Velocity	20% of max. values
Max. Acceleration	20% of max. values
Cartesian planning	True

It also achieved a success rate of 100%. Moreover, it significantly outperformed the RL-based system in terms of insertion speed: the average process time for classical control was only 1.84 s, with a standard deviation of 0.10 s. The histogram in Figure 7 depicts the distribution of process times for both control strategies, clearly showing that the classical control completes the task in less than half the time required by the RL-based control on average.

At the deployment stage, the libfranka library calculates EE forces from measured joint torques. While not highly precise, these calculations are adequate for rough estimations of the applied forces. A comparison of the 3D EE forces exerted by the robot holding the peg during one representative insertion attempt are presented in Figure 8. They reveal the classical control approach to apply significantly less contact force, particularly in the pushing direction x. This method is carefully calibrated to insert the peg without contacting the surface of the hole, thus avoiding jamming. In contrast, the RL control exerts more than double the maximum force—up to 10 N—compared to approximately 5 N for classical control. This indicates that RL control maintains the ability to manage insertion tasks even when not precisely calibrated, making it more adaptable to variations in the environment and minor changes in the setup.

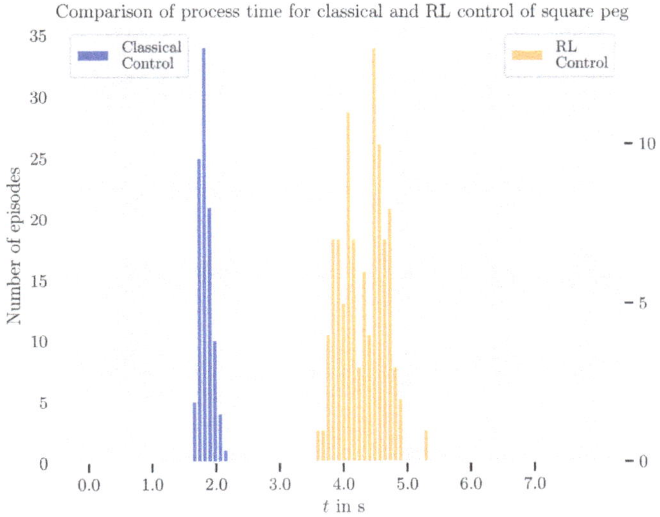

Figure 7. Comparison of process time for the classical and the proposed RL-based control in reality.

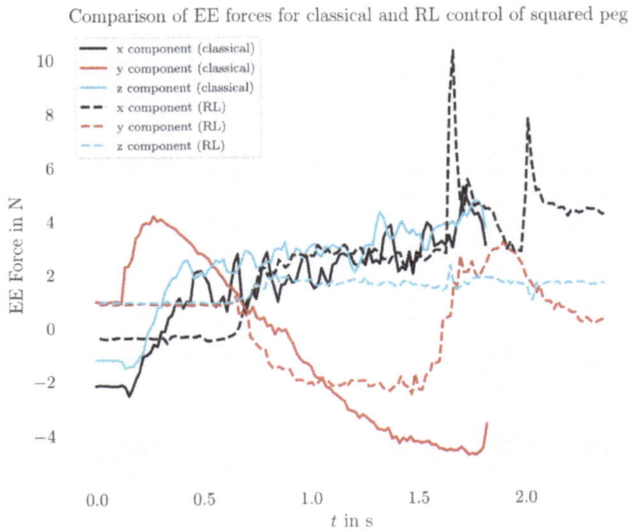

Figure 8. Comparison of EE forces during insertion of a squared peg for the classical and RL control strategy.

5.3. Adapted Peg Shape

We further modified the peg object to have a round shape, with its diameter matching the previous square length. For this variant as well, both control strategies achieved a 100% success rate. The average process time for inserting this round peg applying RL-based control was 3.62 s, with a standard deviation of 0.20 s. The classical control approach achieved an average cycle time of 1.78 s with a standard deviation of 0.09 s. The histogram in Figure 9 shows the distribution of the process times for both control strategies. Once again, the classical control significantly outperformed the RL-based system in terms of cycle time, requiring less than half as long on average.

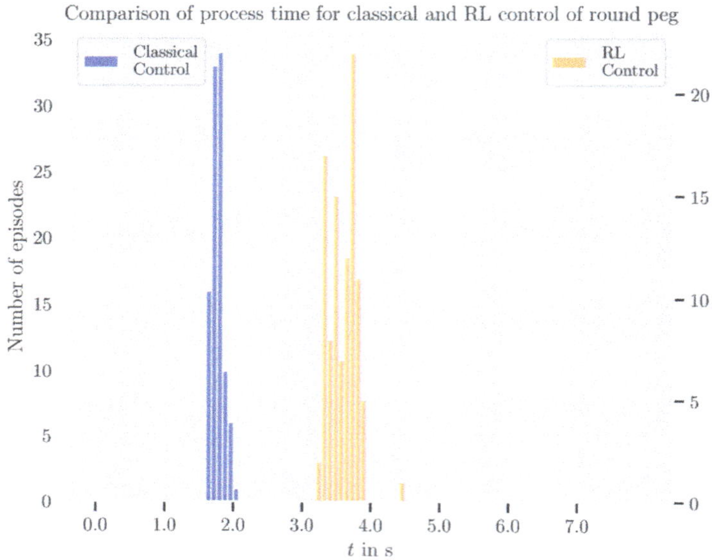

Figure 9. Adaptation of the peg to a circular base area. Comparison of process time for the classical and the proposed RL-based control system in reality.

5.4. Injected Calibration Offset

We evaluated the robustness of the control system towards calibration offsets of both arms, i.e., a perpendicular deviation of the right robot's basis from the original axis introduced in software. The classical trajectory control failed to complete the task in 100% of the cases for an offset bigger than 1.5 mm. This motivates to examine the RL-based control system's performance under similar conditions for a maximum episode duration of 10 s. The histogram in Figure 10 illustrates the distribution of process cycle times for both an offset of 2 mm and 3 mm.

At a shift of 2 mm, the RL approach still achieves a success rate of 47%, while for an offset of 3 mm, the success rate even increases to 63%. However, the average process time increases from 7.64 s to 8.64 s between both setups.

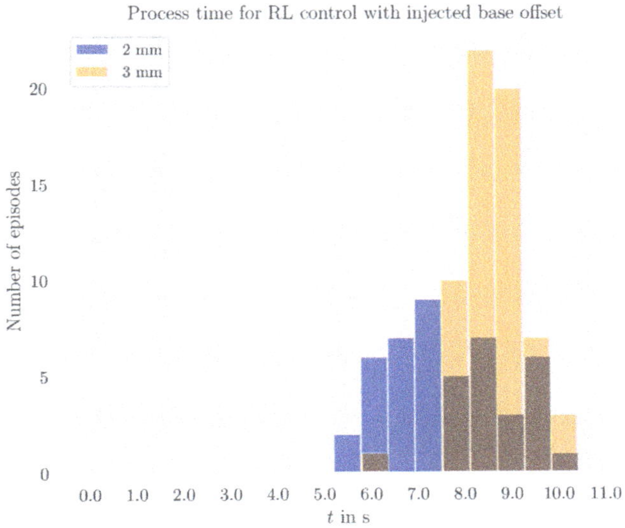

Figure 10. Distribution of process cycle times for RL control with an injected base offset of 2 mm and 3 mm.

6. Discussion

The selection of R_{min} for the curriculum implicitly predetermines the average process time by influencing the convergence of the return R toward this threshold. The maximum episode length parameter, here set to 150 steps, establishes a maximum process duration equivalent to 7.5 s. The simulation reveals a substantial variance in the normally distributed insertion duration, determined by the difference between R_{min} influencing the time needed to accumulate this reward and the maximum episode length. This variance can be reduced, albeit at the expense of a more complex learning process, by narrowing the gap between these two values. This finding is pivotal, as it indicates that for cyclic assembly processes, crucial characteristics can be statistically determined a priori by the curriculum parameters.

The classical control reaches a success rate of 100% in reality and so does our RL-based approach. We expected this, given that the clearance of both peg and hole are 1 mm and thereby one order of magnitude above the Franka's repeatability of pose and path of 0.1 mm. To rank this result in the state of research, we refer to [16]. They achieved a success rate of less than 70% in reality for a similar task. As we also utilized RL with a hierarchical control system, we identified the curriculum-based training strategy and application of PLAI as the key to our success. Even if classical control performed twice as fast as the RL-based control system regarding the process time, the latter offers several advantages. The RL-based control system is more flexible and could be adapted to strongly differing workpiece geometries with minimal effort of retraining given updated CAD files. On the other hand, the classical control system is more sensitive to variations in the environment that may lead to jamming and not finishing the insertion, requiring careful (re)calibrations. Figure 8 shows that the RL-based control actually has a lot of contact at the beginning of the insertion process and manages to prevent jamming by correcting its forces from feedback of the observations. An experiment introducing an error for one of the robot's base positions proved the classical control to fail at an offset of more than 1.5 mm, while the RL-based control system still achieved a success rate of 63% at an offset of 3 mm.

Experiments with a round peg in the real-world setup confirm the classical control to maintain the same process cycle time compared to the squared peg, as minimal contact and no dynamic behavior changes are observed. Meanwhile, the RL-based policy achieves a shorter cycle time compared to the squared peg scenario. This improvement is due to the elimination of the need for precise orientation, which reduces contacts and the likelihood of jamming situations. It also confirms the ability of the RL approach to solve errors in orientation without jamming.

7. Conclusions

In this study, we build upon our previous work by successfully transferring a curriculum-based RL approach for robotic dual-arm assembly from simulation to reality. We focus on a peg-hole task with a 1.0 mm clearance, employing well-established software libraries such as robosuite for impedance control, MuJoCo for physics simulation, SB3 for RL, and gym for seamless interfacing. ROS 2 connects the robots to a workstation that runs the supervisory control policy. The control system architecture employs a low-level impedance controller on each robot, receiving joint position setpoints from the policy.

The training in simulation follows a reverse curriculum strategy, starting each iteration with fulfilling basic return thresholds and progressing to fine-tuning afterwards. We introduce DR in the form of noisy joint angles at initialization to enhance the system's adaptability to various initial states. Post-training, we analyze inferences from the controlled system to assess reliability and average process time, confirming that the curriculum parameters can effectively predict these outcomes, even after deployment on the real-world setup. Traditional control algorithms show less variance and shorter process times compared to our RL approach. Nevertheless, training a control system in simulation enables the system to be confronted with a wide distribution of situations and thereby increases adaptability and flexibility particularly demonstrated for calibration errors.

Currently, high training durations of more than 10 hours on a state-of-the-art desktop PC are limitting the deployment for fast changing requirements. We also require human expertise to manually construct a specific curriculum. This procedure has to be automated. As we first explored capabilities of such Sim2Real transfers, we decided to employ a large clearance of 1 mm. The range of relevant assembly applications reaches 0.5 mm at the upper bound of ISO 286.

Our future work will investigate the effects of bootstrapping a policy leveraging imitation learning from a small number of expert demonstrations to reduce the training time. We plan to explore tighter clearances and different geometries to enhance applicability to relevant real-world assembly tasks. Additionally, integrating low-cost cameras into the setup will allow us to increase adaptivity to task variations, thereby expanding the system's flexibility.

Author Contributions: Conceptualization, K.W., O.D., T.W. and U.F.; methodology, K.W., O.D. and U.F.; software, K.W., R.L. and T.W.; validation, K.W., S.Z. and R.L.; investigation, K.W., S.Z. and R.L.; data curation, S.Z. and R.L.; writing—original draft preparation, K.W.; writing—review and editing, S.Z.; visualization, K.W.; supervision, T.W. and U.F. All authors have read and agreed to the published version of the manuscript.

Funding: This work was funded by Project "Rob4Print-Digital robot training to minimize waste in 3D printing", Federal Ministry for the Environment, Nature Conservation, Nuclear Safety and Consumer Protection (BMUV) Germany through the program DigiRess—Digital applications to increase resource efficiency in circular production processes.

Data Availability Statement: Data are contained within the article.

Conflicts of Interest: The authors declare no conflicts of interest. The funders had no role in the design of the study; in the collection, analyses or interpretation of data; in the writing of the manuscript or in the decision to publish the results.

Abbreviations

The following abbreviations are used in this manuscript:

EE	end-effector
DR	domain randomization

HER	Hindsight Experience Replay
MDPI	Multidisciplinary Digital Publishing Institute
NN	neural network
PLAI	policy-level action integration
RL	reinforcement learning
Sim2Real	simulation-to-reality
SAC	Soft Actor-Critic
SB3	Stable Baselines3

References

1. Wrede, K.; Donath, O.; Wohlfahrt, T.; Feldmann, U. Curriculum-Organized Reinforcement Learning for Robotic Dual-arm Assembly. In Proceedings of the Automation, Robotics & Communications for Industry 4.0/5.0, Innsbruck, Austria, 7–9 February 2024; pp. 8–14. [CrossRef]
2. Jiang, J.; Huang, Z.; Bi, Z.; Ma, X.; Yu, G. State-of-the-Art Control Strategies for Robotic PiH Assembly. *Robot. Comput.-Integr. Manuf.* **2020**, *65*, 101894. [CrossRef]
3. Yuan, F.; Shen, X.; Wu, J.; Wang, L. Design of Mobile Phone Automatic Assembly System Based on Machine Vision. *J. Phys. Conf. Ser.* **2022**, *2284*, 012012. [CrossRef]
4. Zhu, Y.; Wong, J.; Mandlekar, A.; Martín-Martín, R.; Joshi, A.; Nasiriany, S.; Zhu, Y. Robosuite: A Modular Simulation Framework and Benchmark for Robot Learning. *arXiv* **2022**, arXiv:2009.12293. [CrossRef]
5. Todorov, E.; Erez, T.; Tassa, Y. MuJoCo: A Physics Engine for Model-Based Control. In Proceedings of the 2012 IEEE/RSJ International Conference on Intelligent Robots and Systems, Vilamoura-Algarve, Portugal, 7–12 October 2012; pp. 5026–5033. [CrossRef]
6. Raffin, A.; Hill, A.; Gleave, A.; Kanervisto, A.; Ernestus, M.; Dormann, N. Stable-Baselines3: Reliable Reinforcement Learning Implementations. *J. Mach. Learn. Res.* **2021**, *22*, 1–8.
7. Salvato, E.; Fenu, G.; Medvet, E.; Pellegrino, F.A. Crossing the Reality Gap: A Survey on Sim-to-Real Transferability of Robot Controllers in Reinforcement Learning. *IEEE Access* **2021**, *9*, 153171–153187. [CrossRef]
8. Brockman, G.; Cheung, V.; Pettersson, L.; Schneider, J.; Schulman, J.; Tang, J.; Zaremba, W. OpenAI Gym. *arXiv* **2016**, arXiv:1606.01540. [CrossRef]
9. Sutton, R.S.; Barto, A. *Reinforcement Learning: An Introduction*, nachdruck ed.; Adaptive Computation and Machine Learning; The MIT Press: Cambridge, MA, USA, 2014.
10. Zhao, W.; Queralta, J.P.; Westerlund, T. Sim-to-Real Transfer in Deep Reinforcement Learning for Robotics: A Survey. In Proceedings of the 2020 IEEE Symposium Series on Computational Intelligence (SSCI), Canberra, ACT, Australia, 1–4 December 2020; pp. 737–744. [CrossRef]
11. Scheiderer, C.; Dorndorf, N.; Meisen, T. Effects of Domain Randomization on Simulation-to-Reality Transfer of Reinforcement Learning Policies for Industrial Robots. In *Advances in Artificial Intelligence and Applied Cognitive Computing*; Transactions on Computational Science and Computational Intelligence; Arabnia, H.R., Ferens, K., Kozerenko, E.B., Olivas Varela, J.A., Tinetti, F.G., Eds.; Springer International Publishing: Cham, Switzerland, 2021; pp. 157–169. [CrossRef]
12. Tobin, J.; Fong, R.; Ray, A.; Schneider, J.; Zaremba, W.; Abbeel, P. Domain Randomization for Transferring Deep Neural Networks from Simulation to the Real World. In Proceedings of the 2017 IEEE/RSJ International Conference on Intelligent Robots and Systems (IROS), Vancouver, BC, Canada, 24–28 September 2017; pp. 23–30. [CrossRef]
13. Tang, B.; Lin, M.A.; Akinola, I.A.; Handa, A.; Sukhatme, G.S.; Ramos, F.; Fox, D.; Narang, Y.S. IndustReal: Transferring Contact-Rich Assembly Tasks from Simulation to Reality. In Proceedings of the Robotics: Science and Systems XIX, Daegu, Republic of Korea, 10–14 July 2023; Volume 19.
14. Coumans, E.; Bai, Y. PyBullet, a Python Module for Physics Simulation for Games, Robotics and Machine Learning. 2016. Available online: http://pybullet.org (accessed on 19 September 2024) .
15. Paszke, A.; Gross, S.; Massa, F.; Lerer, A.; Bradbury, J.; Chanan, G.; Killeen, T.; Lin, Z.; Gimelshein, N.; Antiga, L.; et al. PyTorch: An Imperative Style, High-Performance Deep Learning Library. In *Advances in Neural Information Processing Systems*; Curran Associates, Inc.: Red Hook, NY, USA, 2019; Volume 32.
16. Alles, M.; Aljalbout, E. Learning to Centralize Dual-Arm Assembly. *Front. Robot. AI* **2022**, *9*, 830007. [CrossRef] [PubMed]
17. Yang, X.; Ji, Z.; Wu, J.; Lai, Y.K. An Open-Source Multi-goal Reinforcement Learning Environment for Robotic Manipulation with Pybullet. In *Towards Autonomous Robotic Systems*; Fox, C., Gao, J., Ghalamzan Esfahani, A., Saaj, M., Hanheide, M., Parsons, S., Eds.; Springer International Publishing: Cham, Switzerland, 2021; pp. 14–24. [CrossRef]
18. Panerati, J.; Zheng, H.; Zhou, S.; Xu, J.; Prorok, A.; Schoellig, A.P. Learning to Fly—A Gym Environment with PyBullet Physics for Reinforcement Learning of Multi-agent Quadcopter Control. In Proceedings of the 2021 IEEE/RSJ International Conference on Intelligent Robots and Systems (IROS), Prague, Czech Republic, 27 September–1 October 2021; pp. 7512–7519. [CrossRef]
19. Mittal, M.; Yu, C.; Yu, Q.; Liu, J.; Rudin, N.; Hoeller, D.; Yuan, J.L.; Singh, R.; Guo, Y.; Mazhar, H.; et al. Orbit: A Unified Simulation Framework for Interactive Robot Learning Environments. *IEEE Robot. Autom. Lett.* **2023**, *8*, 3740–3747. [CrossRef]
20. Serrano-Muñoz, A.; Arana-Arexolaleiba, N.; Chrysostomou, D.; Boegh, S. Skrl: Modular and Flexible Library for Reinforcement Learning. *J. Mach. Learn. Res.* **2022**, *24*, 1–9.

21. Nasiriany, S.; Liu, H.; Zhu, Y. Augmenting Reinforcement Learning with Behavior Primitives for Diverse Manipulation Tasks. In Proceedings of the 2022 International Conference on Robotics and Automation (ICRA), Philadelphia, PA, USA, 23–27 May 2022, pp. 7477–7484. [CrossRef]
22. Wilcox, A.; Balakrishna, A.; Dedieu, J.; Benslimane, W.; Brown, D.; Goldberg, K. Monte Carlo Augmented Actor-Critic for Sparse Reward Deep Reinforcement Learning from Suboptimal Demonstrations. *Adv. Neural Inf. Process. Syst.* **2022**, *35*, 2254–2267.
23. Xu, M.; Veloso, M.; Song, S. ASPiRe: Adaptive Skill Priors for Reinforcement Learning. *Adv. Neural Inf. Process. Syst.* **2022**, *35*, 38600–38613.
24. Park, H.; Bae, J.H.; Park, J.H.; Baeg, M.H.; Park, J. Intuitive Peg-in-Hole Assembly Strategy with a Compliant Manipulator. In Proceedings of the IEEE ISR 2013, Seoul, Republic of Korea, 24–26 October 2013; pp. 1–5. [CrossRef]
25. Broenink, J.F.; Tiernego, M.L.J. Peg-in-Hole Assembly Using Impedance Control with a 6 DOF Robot. In Proceedings of the 8th European Simulation Symposium, Genoa, Italy, 24–26 October 1996; pp. 504–508.
26. Nottensteiner, K.; Stulp, F.; Albu-Schäffer, A. Robust, Locally Guided Peg-in-Hole Using Impedance-Controlled Robots. In Proceedings of the 2020 IEEE International Conference on Robotics and Automation (ICRA), Paris, France, 31 May–31 August 2020; pp. 5771–5777. [CrossRef]
27. Park, H.; Kim, P.K.; Bae, J.H.; Park, J.H.; Baeg, M.H.; Park, J. Dual Arm Peg-in-Hole Assembly with a Programmed Compliant System. In Proceedings of the 2014 11th International Conference on Ubiquitous Robots and Ambient Intelligence (URAI), Kuala Lumpur, Malaysia, 12–15 November 2014; pp. 431–433. [CrossRef]
28. Florensa, C.; Held, D.; Wulfmeier, M.; Zhang, M.; Abbeel, P. Reverse Curriculum Generation for Reinforcement Learning. In Proceedings of the 1st Annual Conference on Robot Learning, PMLR, Mountain View, CA, USA, 13–15 November 2017; pp. 482–495.
29. Aljalbout, E.; Frank, F.; Karl, M. On the Role of the Action Space in Robot Manipulation Learning and Sim-to-Real Transfer. *IEEE Robot. Autom. Lett.* **2024**, *9*, 5895–5902. [CrossRef]
30. Haarnoja, T.; Zhou, A.; Abbeel, P.; Levine, S. Soft Actor-Critic: Off-Policy Maximum Entropy Deep Reinforcement Learning with a Stochastic Actor. In Proceedings of the 35th International Conference on Machine Learning, PMLR, Stockholm, Sweden, 10–15 July 2018; pp. 1861–1870.
31. Andrychowicz, M.; Wolski, F.; Ray, A.; Schneider, J.; Fong, R.; Welinder, P.; McGrew, B.; Tobin, J.; Pieter Abbeel, O.; Zaremba, W. Hindsight Experience Replay. In *Advances in Neural Information Processing Systems*; Curran Associates, Inc.: Red Hook, NY, USA, 2017; Volume 30.
32. Macenski, S.; Foote, T.; Gerkey, B.; Lalancette, C.; Woodall, W. Robot Operating System 2: Design, Architecture, and Uses in the Wild. *Sci. Robot.* **2022**, *7*, eabm6074. [CrossRef] [PubMed]
33. Coleman, D.T.; Sucan, I.A.; Chitta, S.; Correll, N. Reducing the Barrier to Entry of Complex Robotic Software: A MoveIt! Case Study. *arXiv* **2014**, arXiv:1404.3785. [CrossRef]

machines

MDPI

Article

Data-Driven Digital Twin Framework for Predictive Maintenance of Smart Manufacturing Systems

Tarana Khan [1], Urfi Khan [2], Adnan Khan [3], Calahan Mollan [2], Inga Morkvenaite-Vilkonciene [4,5,*] and Vijitashwa Pandey [2,*]

1 Department of Mechanical and Automation Engineering, Indira Gandhi Delhi Technical University for Women, New Delhi 110006, India; tarana012btdmam21@igdtuw.ac.in
2 Industrial & Systems Engineering Department, Oakland University, Rochester, MI 48309, USA; urfikhan@oakland.edu (U.K.); calahanmollan@oakland.edu (C.M.)
3 Institute of Innovation in Technology & Management, New Delhi 110058, India; adnankhan.cs@gmail.com
4 Department of Electrical Engineering, Vilnius Gediminas Technical University, Plytines g. 25, 10105 Vilnius, Lithuania
5 Laboratory of Bioelectrochemical Technologies, Center for Physical Sciences and Technology, Sauletekio 3, 10257 Vilnius, Lithuania
* Correspondence: inga.morkvenaite-vilkonciene@vilniustech.lt (I.M.-V.); pandey2@oakland.edu (V.P.)

Abstract: A Digital twin (DT) enables the acquisition and subsequent analysis of large amounts of process data. Various machine learning (ML) algorithms exist for analysis and prediction that can be used in this scenario. However, there is very little understanding of the relative merit of these methods. This paper proposes a DT framework in the context of predictive maintenance in smart manufacturing to compare the prediction efficacy of prevalent ML models. Data-driven models were developed using machine learning algorithms to predict surface roughness and power consumption during a CNC turning operation. Three process parameters, namely cutting velocity, feed rate, and depth of cut, and two dependent parameters, surface roughness and power consumption, were selected for model development. Seven ML algorithms were tested for each response parameter: Linear Regression, XGB Regressor, Random Forest Regressor, Average Ensemble, AdaBoost Regressor, SVR, and MLP. The results of the comparative analysis of the ML algorithms showed that the Random Forest Regressor is the best prediction model for surface roughness, with the highest R^2 (94.2% ± 2.4%), lowest MAE (0.011 ± 0.002), lowest MAPE (15.6% ± 4.0%), and lowest RMSE (0.017 ± 0.003), while the XGB Regressor demonstrated the best performance for power consumption prediction, with the highest R^2 (98.9% ± 0.5%), lowest MAE (22.513 ± 4.424), lowest MAPE (3.0% ± 0.5%), and lowest RMSE (42.650 ± 8.933). The best-performing machine learning algorithm was subsequently utilized in the data-driven models, helping to achieve an improved surface finish. This enables predictive maintenance, reducing energy usage for more sustainable production.

Keywords: digital twin; smart manufacturing; Industry 4.0; predictive maintenance; cyber-physical production systems; machine learning

1. Introduction

Technological advancements have significantly transformed manufacturing systems under the emerging idea of Industry 4.0 technologies, namely the Internet of Things (IoT), machine learning, and big data analytics [1–3]. Another new evolution in the fourth Industrial Revolution, namely cyber-physical production systems (CPPS), provides critical inputs for semi-autonomous and autonomous decision-making for the future of manufacturing

https://doi.org/10.3390/machines13060481

systems [4,5]. Cyber-physical production systems combine physical manufacturing systems with advanced networking technologies, enabling efficient monitoring and handling of physical processes [6]. CPPS enables real-time, enhanced performance in manufacturing systems that leads to customized product development and resource-efficient production, increasing reliability and resilience in production processes [7].

Integrating these technologies into the existing manufacturing systems enables better decision-making by enhancing overall quality, productivity, and sustainability [8,9]. These technologies enable the manufacturing systems to become 'smart'. Smart systems are knowledge-driven, well-connected, and dynamically adaptive to changes [10]. As a result, manufacturers can achieve higher quality standards and increased production rates with more sustainable practices, which enable them to stay competitive in the rapidly changing industrial scenario. With the increased prevalence and sophistication of various machine learning tools that help with analysis and prediction, deciding which tool to use is often challenging [11].

A major component of CPPS is the digital twin, which is basically a digital counterpart of the physical system. A DT is a virtual model that utilizes real-time data from a physical object to simulate its behavior [12]. Digital twins are very promising in systems where real-time monitoring is essential, as they assist in the supervision and analysis of manufacturing systems, facilitating effective fault diagnosis and process planning [13]. DTs significantly enhance the operator's decision-making capabilities by centralizing and compiling relevant information about the system. They have the potential to accurately analyze the performance of a process or system based on real-world data, such as working conditions and real-time status of physical elements, captured through sensors that are attached to the manufacturing system. To ensure that the digital twin closely aligns with the behavior of the physical system, periodic updates of the process parameters are necessary to minimize the variation between the actual and simulated behavior, thus improving accuracy [14]. These types of 'time-maintained' digital twins represent a key goal for smart manufacturing.

Data-driven digital twinning plays a crucial role in enhancing the reliability and performance of the manufacturing systems. Incorporating machine learning into this digital twin improves its functionality and potential. Advanced ML algorithms, trained on huge amounts of operational and real-time sensor data, allow for the precise analysis of system behavior and the accurate prediction of future outcomes [15]. This enables the DT to identify potential failures at an early stage, allowing for timely preventive measures to improve the overall efficiency and real-time performance of manufacturing systems.

Studies have demonstrated the effectiveness of such data-driven digital twinning frameworks across various manufacturing applications. For instance, in the cooling tower systems of automotive production plants, DTs have been utilized to optimize energy consumption and cooling efficiency using polynomial and linear regression techniques [16]. Similarly, in aluminum die-casting manufacturing, a real-time visual quality prediction system using XGBoost and deep learning-based defect detection achieved over 95% defect identification accuracy, reducing false detection rates and improving production reliability [17]. Akinsolu and Zribi [18] introduced a comprehensive framework which integrates 19 different ML algorithms, such as Random Forests, SVR, and Decision Trees, and evaluates and ranks model performance within a continuous-flow manufacturing environment. This framework assists in identifying the most suitable predictors for the accurate modeling of system variables like temperature and pressure [18]. Collectively, these studies highlight the successful application of DT in improving operational visibility, predictive capabilities, and system adaptability in industrial settings.

Combining AI–ML and big data, digital twinning is evolving rapidly, and there are potentially many application possibilities in the industry such as real-time equipment health monitoring, process optimization, and predictive maintenance scheduling in manufacturing systems [19,20]. For example, digital twins can track machine performance to anticipate failures, optimize cutting parameters in machining operations, and improve energy efficiency across production lines. These capabilities allow for real-time monitoring and data-driven decision-making, providing significant benefits in the predictive maintenance of physical systems [21,22]. Physics-based simulation and the digital twin concept can also be used to calculate the Remaining Useful Life (RUL) of manufacturing equipment, offering accurate RUL predictions that support maintenance planning, production scheduling, and operational decision-making processes [23].

Traditional maintenance methods are usually reactive, addressing issues after they occur, leading to increased downtimes, sudden production stops, and operational costs. Predictive Maintenance (PdM) can be more effective than traditional maintenance approaches, such as Reactive Maintenance (RM) and Preventive Maintenance (PM), in the context of Industry 4.0 [24]. PdM is a strategy used to forecast equipment failures, enabling preventive actions and maintenance to be planned before actual failure [25–27]. PdM has the potential to reduce unexpected downtimes and extend the lifespan of equipment, which in turn improves the efficiency and reliability of the physical system [28,29]. This approach uses inspection and diagnosis to detect anomalies in the equipment, facilitating early preventive actions [30,31]. In this way, PdM transforms unexpected failures into planned repairs, reducing the chances of production downtime and associated costs.

In some cases, simple regression models can be used for predictive modeling [32]. For example, different regression models based on correlation analyses can be used to effectively predict orbital welding time in single-item production systems. However, their effectiveness highly depends on process knowledge and the standardization of data.

The integration of DT technology with ML offers significant potential to enhance smart manufacturing systems, particularly in the field of predictive maintenance [33,34]. Although numerous ML algorithms exist for predictive maintenance, there is limited literature comparing their performance when applied within a DT framework for manufacturing processes [35]. This highlights a key research gap, as selecting the most appropriate ML algorithm is essential to ensuring accurate and reliable DT-based predictions. To aid algorithm selection, this study proposes and utilizes a data-driven digital twin framework to evaluate and compare the relative performance of various ML algorithms in predicting critical response parameters, particularly surface roughness and power consumption during CNC turning operations. The study aims to identify the most effective model for incorporation into a DT environment that enables real-time monitoring and preventive maintenance. The application area relates to a CNC turning process to predict energy usage and surface roughness to achieve more sustainable and cost-effective production. In the context of this application, an accurate prediction model will help achieve a better surface finish for high-quality and customized products.

2. Methods

2.1. Framework Development

2.1.1. Maintenance System Architecture

An architectural overview of the proposed digital twin framework, focused on predictive maintenance, is illustrated in Figure 1. This framework consists of two cycles, one for comparing and selecting the algorithms and the other for the implementation part, where the operator performs decision-making regarding maintenance, preventive actions, and process parameter control. In the first cycle, the data gathered through sensors attached to the physical system (a CNC turning machine in the present case) undergo data pre-processing to prepare it for implementing ML algorithms. These data are used to train and test various ML algorithms. The best algorithms that provide the most accurate predictions are utilized in the data-driven models to predict key response parameters. This may assist the operator in the predictive maintenance of manufacturing systems for improved performance and decision-making. Additionally, if another algorithm starts predicting better later in the system lifecycle, possibly because of user inputs, system degradation, etc., it becomes the algorithm of choice. The framework suggests that the system periodically evaluates the performance of multiple machine learning algorithms as new data are collected during operation. This process can be automated by using performance metrics to assess the effectiveness of each algorithm on the updated dataset. When a different algorithm consistently outperforms the current model based on the predefined metrics, the operator can switch to the parameter values suggested by the now better-performing algorithm. This adaptive mechanism enables continuous optimization of the system's predictive capability over its lifecycle. Thus, this framework effectively assists in providing feedback to the operators, enabling them to take preventive actions on the physical system before any actual failure occurs. This subsequently allows for the continuous optimization of process parameters throughout the manufacturing process, resulting in sustained prediction accuracy despite process drift.

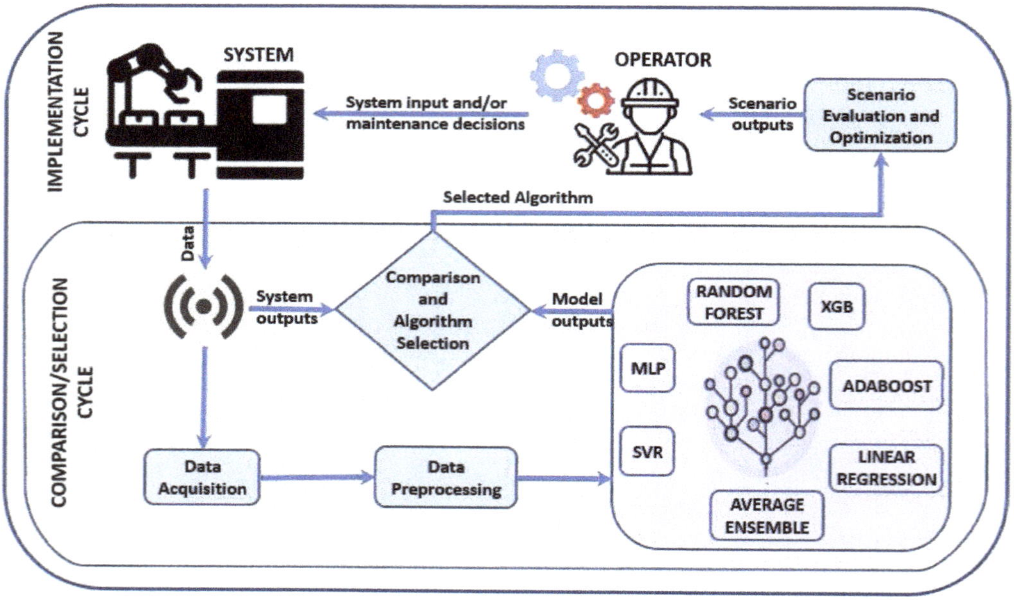

Figure 1. Proposed Data-Driven Digital Twin Framework.

2.1.2. Implementation of the Proposed System Architecture

The framework of the proposed system offers a sequential procedure for developing a data-driven machine learning-based DT model for predicting critical response parameters. This framework involves four stages: data collection, data pre-processing, training and testing ML models, and selecting the best predictive models.

In this study, we have utilized the dataset based on the CNC turning process, derived from experimental observations conducted by Andre Dorigueto Canal and Andrerson Vincente Borille at the Competence Center in Manufacturing (CCM), a laboratory within the

Aeronautics Institute of Technology (ITA), Brazil [36]. The dataset provided readings on surface roughness and force measurements that assist in evaluating machining performance.

These experiments were conducted on AISI H13 steel (mean hardness = 200 HV) using a ROMI E280 CNC lathe with a maximum rotational speed of 4000 rpm and a nominal power of 18.5 kW. A Sandvik Coromant cutting tool (ISO TNMG 16 04 04-PE 4425) with a tool shank (ISO MTJNL 2020K 16M1) was used for the machining process. Surface roughness was measured with a Mitutoyo Surftest (SJ-210) portable roughness tester. Additionally, cutting forces were measured using a Kistler Type (9265B) dynamometer connected to a Kistler Type charge amplifier and operated through Kistler Dynoware (2825A) data acquisition software.

After gathering the required data, the dataset was pre-processed to prepare it for use in machine learning models. Data pre-processing involved two steps: data cleaning and feature engineering.

Data Cleaning: The dataset was comprised of 27 columns and 324 samples. Two columns that contained inconsistent and invalid data were removed. These fields were "Run_ID", an experimental identifier without any analytical relevance, and "C_TIME", which had excessive missing values. All other columns were numeric type except for "Position" which was a categorical variable describing work piece placement. To enable numerical analysis, one-hot encoding was applied to "Position", converting the categorical variable into a binary representation. Once all features were numeric, a correlation matrix was generated using the *matplotlib* and *seaborn* libraries to see which features correlated strongly with surface roughness. All of the features with a correlation coefficient between −0.5 and 0.5 were dropped from the dataset. This was subsequently analyzed using expert domain knowledge to identify the most relevant process parameters that effectively map with the response parameters and significantly influence the output or response parameters to be predicted. Based on the analysis, a normalized list of five key process parameters (velocity, feed, and depth of cut for roughness; force and velocity for power) was selected to train and test the ML algorithms.

Additionally, hyperparameter tuning was performed for each ML algorithm using Ridge Regression, Grid Search, and Randomized Search methods to identify the hyperparameters that maximize model performance. The results of the hyperparameter tuning, including the best parameters for each model, are illustrated in Table 1 for surface roughness prediction and Table 2 for surface power consumption prediction.

Table 1. ML Hyperparameter Tuning Methods and Best Parameters for Models in Surface Roughness Prediction.

Model	Tuning Method	Best Parameters
Linear Regression	Ridge Regression	{'alpha': 1, 'fit_intercept': True, 'solver': 'auto'}
XGB Regressor	Randomized Search	{'colsample_bytree': 0.78, 'gamma': 0.007, 'learning_rate': 0.29, 'max_depth': 8, 'n_estimators': 341, 'subsample': 0.75}
Random Forest Regressor	Grid Search	{'max_depth': None, 'max_features': 'log2', 'min_samples_leaf': 1, 'min_samples_split': 5, 'n_estimators': 100}
AdaBoost Regressor	Grid Search	{'estimator__max_depth': 3, 'learning_rate': 1.0, 'loss': 'square', 'n_estimators': 50
SVR	Randomized Search	{'C': 1.03, 'epsilon': 0.105, 'gamma': 'scale', 'kernel': 'rbf'
MLP	Randomized Search	{'activation': 'tanh', 'alpha': 0.00208, 'batch_size': 128, 'hidden_layer_sizes': (50, 50), 'learning_rate_init': 0.0619}

Table 2. ML Hyperparameter Tuning Methods and Best Parameters for Models in Power Consumption Prediction.

Model	Tuning Method	Best Parameters
Linear Regression	Ridge Regression	{'alpha': 0.01, 'fit_intercept': True, 'solver': 'svd'}
XGB Regressor	Randomized Search	{learning_rate = 0.29, max_depth = 8, n_estimators = 341, subsample = 0.75, colsample_bytree = 0.78, gamma = 0.007}
Random Forest Regressor	Grid Search	{'estimator__max_depth': 3, 'learning_rate': 1.0, 'loss': 'square', 'n_estimators': 200}
AdaBoost Regressor	Grid Search	{'estimator__max_depth': 3, 'learning_rate': 1.0, 'loss': 'square', 'n_estimators': 50}
SVR	Randomized Search	{'C': 98.96484985318548, 'epsilon': 0.03416347415306914, 'gamma': 'auto', 'kernel': 'rbf'}
MLP	Randomized Search	{'activation': 'logistic', 'alpha': 0.010151348978007608, 'batch_size': 32, 'hidden_layer_sizes': (50, 50), 'learning_rate_init': 0.015226341829186325}

Out of the two target response parameters, the first one, the surface roughness of the machined workpiece, was already present in the dataset. The second response parameter, i.e., power consumption, was absent from the dataset. Fortunately, the dataset had values for cutting force and velocity. This assisted in estimating the missing parameter indirectly by using the available features.

Feature Engineering: Feature engineering was used to transform the available data into more meaningful features that enhanced the overall efficiency of the predictive model by using the fundamental relationships between input and output variables.

One such critical feature introduced was power consumption (P), a critical parameter for machine maintenance. Since power consumption is directly influenced by the cutting force (F) and the cutting velocity (V) during a machining operation, it was derived using the fundamental physical principle of mechanical power in translational motion expressed in Equation (1):

$$P = FV \tag{1}$$

The above relationship has been derived from Newtonian mechanics, where power is defined as the rate of work done, and work as the product of force and displacement. For a constant force applied in the direction of motion along with some displacement over time, power can be expressed as the product of force and velocity. This formulation has been used to estimate power consumption during machining operation.

Apart from this, no additional feature engineering steps were performed in this study.

2.1.3. Training and Testing of ML Models

After acquiring the relevant process parameters, the prediction models were selected, trained and evaluated. Various machine learning algorithms were used for evaluation. These ML algorithms were deployed in Python 3.13.3, with 80% of the dataset used for training and the remaining 20% for evaluating the accuracy on unseen data. This training and testing procedure was subsequently repeated for each ML model to determine their accuracy rates.

The performance of these algorithms was evaluated, and the best-performing ones were identified for potential use in the data-driven predictive models. These models enable precise predictions that support the predictive maintenance of manufacturing systems, leading to improved performance and better decision-making.

2.2. Data-Driven Predictive Models

The study focused on training data-driven predictive models using machine learning algorithms. Seven machine learning algorithms were selected for evaluation: Linear Regression, the XGB Regressor, the Random Forest Regressor, Average Ensemble, the AdaBoost Regressor, SVR, and MLP. The training and testing of all these models was done using the Scikit-learn Python library.

Models were selected which represent the whole spectrum of the machine learning models. The selected models are widely recognized in the literature for their prediction accuracy. Ensemble algorithms, namely the XGB (extreme gradient boosting) Regressor, the Random Forest Regressor, Average Ensemble, and the AdaBoost (adaptive boosting) Regressor, were selected due to their high prediction power and robustness relative to other algorithms [37]. Neural network algorithms, MLP (multilayer perceptron), and SVR (support vector regression) were evaluated due to their prevalence and universal approximation. Linear regression was chosen as a baseline model and for its interpretability [38]. Random Forest was selected as a complex and non-linear model. SVR represents a kernel-based flexible prediction model. XGBoost and AdaBoost were selected due to their high prediction power and robustness relative to other algorithms.

These ML algorithms were evaluated to test the effectiveness of prediction for surface roughness and power consumption. Based on the results of the evaluation, two individual predictive models, Random Forest and XGBoost, were identified as the most suitable ML algorithms due to their accuracy of prediction.

The first model was utilized to predict surface roughness, an important parameter of workpieces for high-quality and reliable products. This model used depth of cut, cutting speed, and feed rate as input parameters to successfully predict the surface roughness using machine learning algorithms.

The second model was used in the prediction of power consumption, which is crucial for cost reduction and energy-efficient production. The related data fields obtained from the dataset have not shown any correlation with power consumption. Thus, based on expert knowledge, feature engineering was used to derive power consumption using cutting force and cutting speed as input parameters, to build a meaningful predictive model that assists in the successful prediction of power consumption.

The data-driven predictive models thus formed assist in the predictive maintenance of the system in achieving a high surface finish with minimum energy usage for more efficient and economical production.

3. Results

After the training phase, the accuracy rate of each ML algorithm was analyzed by comparing the predicted value with the actual value of measurements. For the purpose of evaluating the prediction performance of every algorithm, various statistical measures, namely R^2 (R-squared), MAE (Mean Absolute Error), RMSE (Root Mean Square Error), and MAPE (Mean Absolute Percentage Error), were utilized as illustrated in Tables 1 and 2.

R^2 indicates the quantity of total variation in the output variable which the model managed to reduce. A higher R^2 value signifies a better fit for the model [39].

R^2 is calculated as in Equation (2):

$$R^2 = 1 - \frac{\sum_{i=1}^{n}(yi - \hat{y}_i)^2}{\sum_{i=1}^{n}(yi - \overline{y})^2} \quad (2)$$

MAE is a measure of the absolute value of the average difference between the predictions and the actual failures in the dataset. A small MAE value indicates better predictions for failures [40].

The formula for MAE is as shown in Equation (3):

$$MAE = \frac{\sum_{i=1}^{n} |y_i - \hat{y}_i|}{n} \quad (3)$$

In the above equation, $\hat{y}_i$ is the predicted value and y_i is the actual value.

RMSE represents the standard deviation of the prediction errors. A model's prediction is considered more accurate if the RSME value turns out to be smaller [41].

RMSE is calculated as shown in Equation (4):

$$RMSE = \sqrt{\frac{\sum_{i=1}^{n} (y_i - \hat{y}_i)^2}{n}} \quad (4)$$

MAPE is the mean of absolute percentage errors. Unlike MAE, MAPE measures the variation in percentage compared to the absolute values [42].

MAPE is calculated as shown in Equation (5):

$$MAPE = \frac{1}{n} \sum_{i=1}^{n} \left| \frac{A_i - P_i}{A_i} \right| \quad (5)$$

In the above equation, A_i denotes the i-th actual value, P_i denotes the i-th predicted value, and n denotes the number of data points.

For surface roughness prediction, the performance results of ML algorithms were evaluated as illustrated in Table 3. As ML algorithms are stochastic in nature, each model was tested over 30 independent runs. To ensure the robustness and reliability of the evaluation, descriptive statistics including the mean and standard deviation of the performance metrics (MAE, MAPE, RMSE, and R²), were recorded. Based on the results of evaluation, the Random Forest Regressor achieved the best performance on the test dataset with the highest R^2 value (94.2% ± 2.4%), lowest MAE value (0.011 ± 0.002), lowest MAPE value (15.6% ± 4.0%), and lowest RMSE value (0.017 ± 0.003).

Table 3. ML Algorithm Accuracy for Surface Roughness Prediction (Mean ± Standard Deviation).

ML Algorithm	MAE	MAPE	RMSE	R^2
Linear Regression	0.020 ± 0.002	26.5% ± 3.4%	0.027 ± 0.003	86.0% ± 2.8%
XGB Regressor	0.020 ± 0.003	24.6% ± 4.8%	0.027 ± 0.004	85.0% ± 5.3%
Random Forest Regressor	0.011 ± 0.002	15.6% ± 4.0%	0.017 ± 0.003	94.2% ± 2.4%
Average Ensemble	0.016 ± 0.001	23.0% ± 3.8%	0.018 ± 0.002	93.1% ± 2.1%
AdaBoost Regressor	0.059 ± 0.003	56.2% ± 4.6%	0.067 ± 0.004	9.7% ± 18.9%
SVR	0.022 ± 0.005	29.4% ± 8.0%	0.027 ± 0.006	84.5% ± 8.5%
MLP	0.018 ± 0.002	24.7% ± 4.1%	0.023 ± 0.003	89.4% ± 2.0%

Additionally, a statistical significance analysis was performed to compare the Random Forest Regressor with other models (as illustrated in Figure 2). The results of this analysis, including the Z-scores and p-values, are illustrated in Table 4. The analysis showed that the Random Forest Regressor significantly outperformed other models, as indicated by low p-values (<0.05) for comparisons with Linear Regression, the XGB Regressor, Average Ensemble, SVR, and MLP (as illustrated in Table 4). The AdaBoost Regressor, however, did not show statistically significant improvement when compared to the Random Forest Regressor.

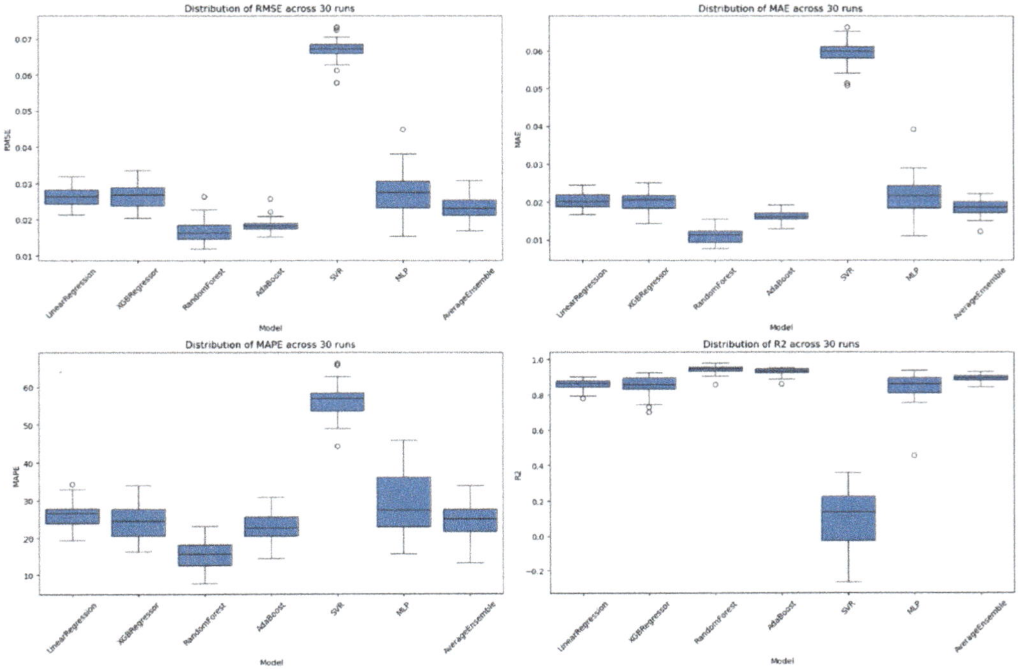

Figure 2. Distribution of Performance Metrics Across 30 Runs for All ML Algorithms in Surface Roughness Prediction.

Table 4. Statistical Significance Analysis Comparing the Random Forest Regressor with Other Models for Surface Roughness Prediction.

Comparison	Z-Score	*p*-Value	Significant ($p < 0.05$)
Random Forest vs. Linear Regression	12.0094	2.28606×10^{-17}	True
Random Forest vs. XGB Regressor	8.25095	2.32836×10^{-11}	True
Random Forest vs. Average Ensemble	8.20355	2.7945×10^{-11}	True
Random Forest vs. AdaBoost	1.95375	0.055559	False
Random Forest vs. SVR	23.9245	9.76039×10^{-32}	True
Random Forest vs. MLP	5.9513	1.63957×10^{-7}	True

Moreover, the Q–Q (quantile–quantile) plot was also used to compare any deviation between the actual and predicted values in the various ML algorithms. A Q–Q plot is a graph depicting the quantiles of actual values set against the corresponding quantiles of predicted values (as illustrated in Figure 4). A perfect prediction would be represented by all the points on the plot following a linear pattern; any differences in the actual and predicted distributions are represented by non-linear relationships [43].

Based on the comparative analysis of the statistical metrics and the Q–Q plot results, the Random Forest Regressor was identified as the best ML algorithm for surface roughness prediction. In contrast, the AdaBoost Regressor showed the worst performance and proved insufficient in the prediction process.

Figures 3 and 4 illustrate the comparative performance of all seven algorithms based on the test dataset. Undoubtedly, as shown by the visual representations, the Random Forest Regressor achieved better prediction results than other algorithms.

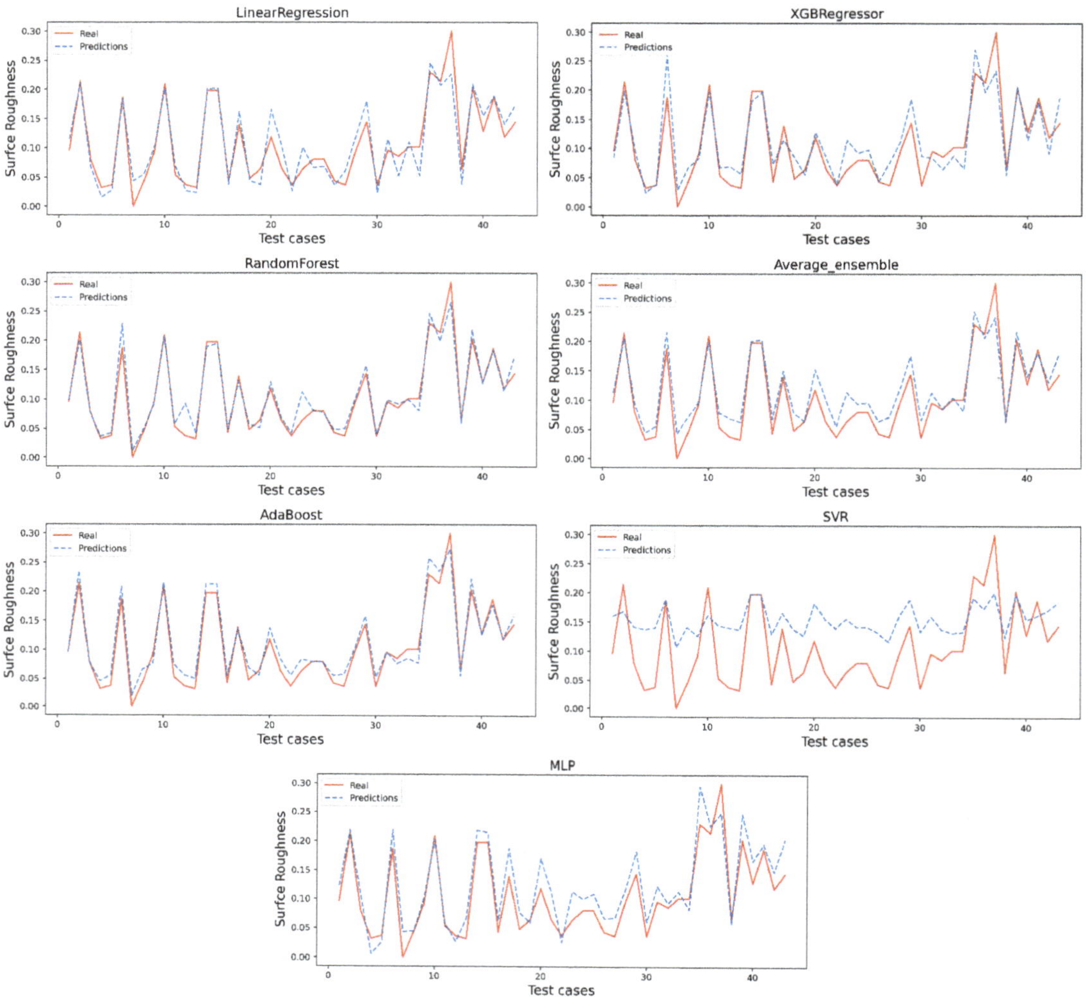

Figure 3. Comparison of ML Algorithm Results: Actual vs. Predicted Surface Roughness Values.

Similarly, for power consumption prediction (as illustrated in Table 5), the best results were obtained for the XGB Regressor in the test dataset, with the highest R^2 value (98.9% ± 0.5%), lowest MAE value (22.513 ± 4.424), lowest MAPE value (3.0% ± 0.5%), and lowest RMSE value (42.650 ± 8.933).

Table 5. ML Algorithm Accuracy for Power Consumption Prediction (Mean ± Standard Deviation).

ML Algorithm	MAE	MAPE	RMSE	R^2
Linear Regression	64.076 ± 3.503	11.5% ± 1.2%	87.650 ± 4.477	95.2% ± 0.4%
XGB Regressor	22.513 ± 4.424	3.0% ± 0.5%	42.650 ± 8.933	98.9% ± 0.5%
Random Forest Regressor	34.959 ± 4.336	4.7% ± 0.6%	52.650 ± 7.886	98.3% ± 0.5%
Average Ensemble	49.369 ± 3.001	7.8% ± 0.5%	58.650 ± 3.119	97.9% ± 0.3%
AdaBoost Regressor	29.339 ± 4.861	3.8% ± 0.8%	57.650 ± 11.564	97.9% ± 0.7%
SVR	75.970 ± 12.434	6.3% ± 1.0%	186.650 ± 21.157	78.5% ± 3.5%
MLP	34.231 ± 4.613	3.8% ± 0.5%	58.650 ± 8.667	97.8% ± 0.5%

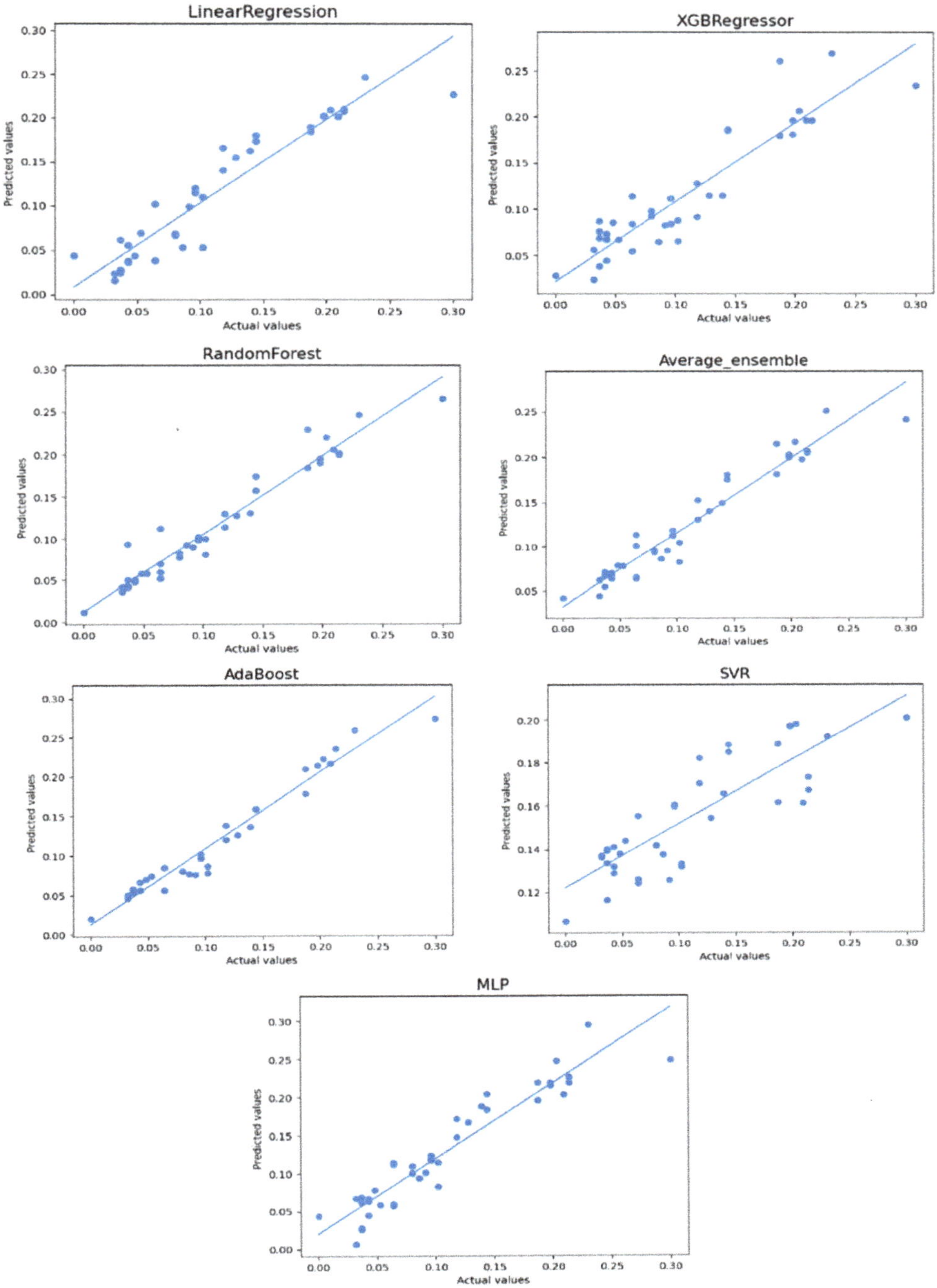

Figure 4. Q–Q Plot: Actual vs. Predicted Surface Roughness Values.

The statistical significance analysis (as illustrated in Figure 5) performed to compare the XGB Regressor with other models showed that the XGB Regressor significantly outperformed other models, as indicated by low p-values (<0.05) for comparisons with Linear

Regression, the Random Forest Regressor, Average Ensemble, the AdaBoost Regressor, SVR, and MLP (as illustrated in Table 6).

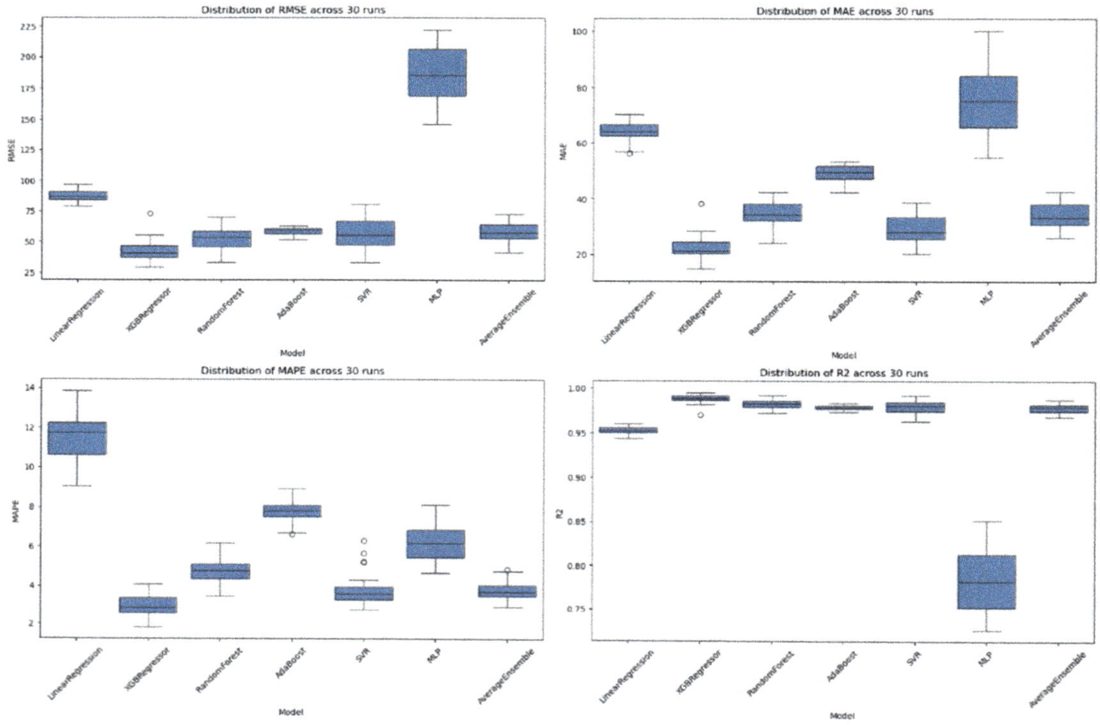

Figure 5. Distribution of Performance Metrics Across 30 Runs for All ML Algorithms in Terms of Power Consumption Prediction.

Table 6. Statistical Significance Analysis Comparing XGB Regressor with Other Models for Power Consumption Prediction.

Comparison	Z-Score	*p*-Value	Significant ($p < 0.05$)
XGB Regressor vs. Linear Regression	29.9667	5.46433×10^{-37}	True
XGB Regressor vs. Random Forest	4.6573	1.914×10^{-5}	True
XGB Regressor vs. Average Ensemble	8.20355	2.7945×10^{-11}	True
XGB Regressor vs. AdaBoost	7.78024	1.09453×10^{-13}	True
XGB Regressor vs. SVR	5.63292	5.44446×10^{-7}	True
XGB Regressor vs. MLP	31.2461	5.55743×10^{-38}	True

Moreover, the XGB Regressor also showed the best Q–Q plot results among all seven algorithms (as illustrated in Figure 7). Although the Random Forest Regressor had a very close R^2 value (98.3% $\pm$ 0.5%), it slightly lagged behind the XGB Regressor. In contrast, the SVR algorithm showed the worst performance and proved insufficient for the prediction process.

Figures 6 and 7 illustrate the comparative performance of all seven algorithms based on the test dataset. Undoubtedly, as shown by the visual representation, the XGB Regressor achieved better prediction results than other algorithms.

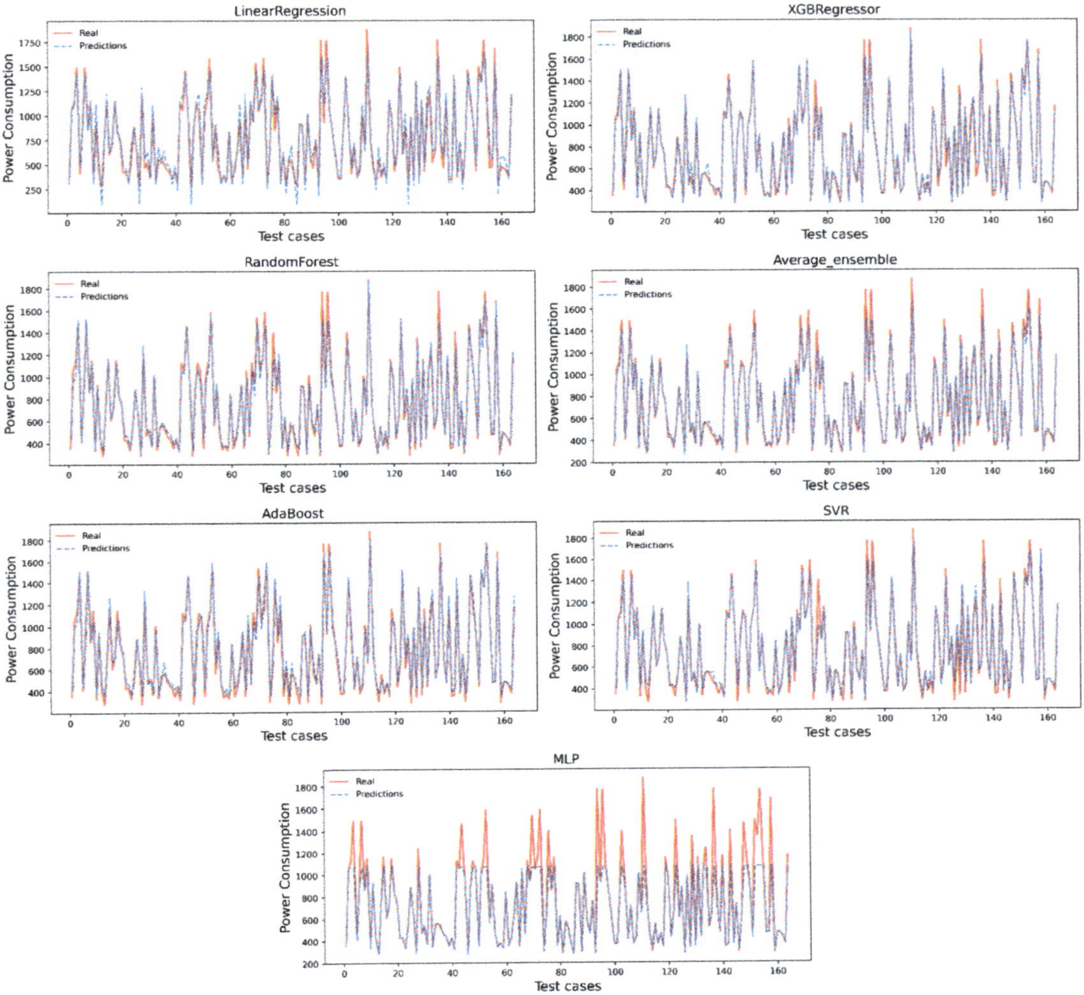

Figure 6. Comparison of ML Algorithm Results: Actual vs. Predicted Power Consumption Values.

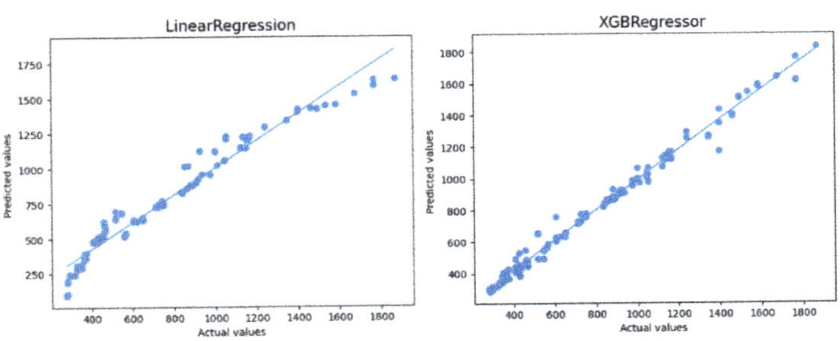

Figure 7. *Cont.*

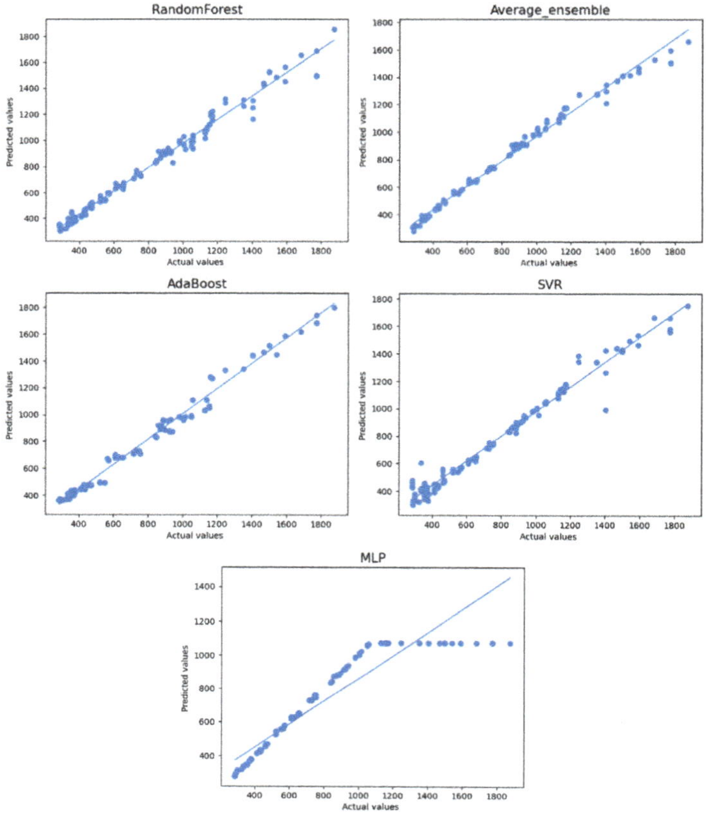

Figure 7. Q–Q Plot: Actual vs. Predicted Power Consumption Values.

The results indicate that the Random Forest Regressor is the best surface roughness prediction model while the XGB Regressor is the best power consumption prediction model. Implementing these algorithms into the predictive models will assist in effective predictive maintenance by reducing energy usage and improving surface finish to achieve more sustainable and effective production. This way, the data-driven predictive models act as a digital twin, enabling better decision-making by enhancing the overall quality, productivity, and sustainability of the existing production system.

4. Discussion

This paper proposes a digital twin framework for smart manufacturing systems to compare different ML analysis techniques for predictive maintenance. The evaluation results show the effectiveness of the proposed framework in predicting surface roughness and power consumption during CNC turning operations. The framework offers a thorough approach for developing and comparing machine learning-based digital twin models for predictive maintenance in manufacturing systems, allowing for prediction and actionable feedback to operators for preventive actions. This enables predictive maintenance of the turning process by minimizing energy usage and assists in producing high-quality products with good surface finish, leading to more sustainable and cost-effective production. The findings indicate that among the seven ML models tested, the Random Forest Regressor achieved the best performance for surface roughness prediction, with the highest R^2 (94.2% ± 2.4%), lowest MAE (0.011 ± 0.002), lowest MAPE (15.6% ± 4.0%), and lowest

RMSE (0.017 ± 0.003), while the XGB Regressor achieved the best performance for power consumption prediction, with the highest R^2 (98.9% ± 0.5%), lowest MAE (22.513 ± 4.424), lowest MAPE (3.0% ± 0.5%), and lowest RMSE (42.650 ± 8.933).

It is important to acknowledge that the current study is based on static experimental data and uses standard regression-based machine learning models, which do not capture temporal dependencies inherent in dynamic manufacturing environments. Future work should address this limitation by incorporating real-time data acquisition systems and time-series databases to develop more adaptive, temporally-aware predictive models. Additionally, future studies should consider applying the proposed framework to diverse, real-world industrial datasets, such as high-volume automotive parts machining, aerospace component manufacturing, or energy-efficient metal cutting applications, in order to further validate the models in different production settings.

Author Contributions: Conceptualization: U.K., T.K., I.M.-V. and V.P.; Methodology: A.K., T.K. and C.M.; Software: T.K. and A.K.; Validation: U.K. and A.K.; Formal analysis: U.K., V.P. and C.M.; Investigation: T.K. and A.K.; Data curation: T.K. and U.K.; Writing: all authors; Visualization: T.K. and V.P.; Supervision: U.K., V.P. and I.M.-V.; Project Administration: V.P. and I.M.-V. All authors have read and agreed to the published version of the manuscript.

Funding: Inga Morkvenaite-Vilkonciene's research was funded by the Research Council of Lithuania (LMTLT), agreement No S-MIP-22–87.

Data Availability Statement: These data were derived from the following resources available in the public domain: CNC turning: roughness, forces and tool wear, https://doi/10.34740/KAGGLE/DS/2205074 (accessed on 28 March 2025).

Acknowledgments: We would like to express our sincere appreciation to the authors (Andre Dorigueto Canal and Andrerson Vincente Borille) of the dataset used in this study which was obtained from Kaggle [28]. Their work provided valuable data that helped in our research.

Conflicts of Interest: The authors declare no conflicts of interest.

References

1. Javaid, M.; Haleem, A.; Singh, R.P.; Suman, R.; Gonzalez, E.S. Understanding the adoption of Industry 4.0 technologies in improving environmental sustainability. *Sustain. Oper. Comput.* **2022**, *3*, 203–217. [CrossRef]
2. Hulbert, S.; Mollan, C.; Pandey, V. Fault Diagnosis and Prediction in Automotive Systems with Real-Time Data Using Machine Learning. *SAE Tech. Pap. 2022-01-0217* **2022**. [CrossRef]
3. Slon, C.; Pandey, V.; Kassoumeh, S. Mixture distributions in autonomous decision-making for industry 4.0. *SAE Int. J. Mater. Manuf.* **2019**, *12*, 135–148. [CrossRef]
4. SCOOP [Internet]. Tessenderlo, Limburg. 2022. Available online: https://www.i-scoop.eu/industry-4-0/ (accessed on 3 July 2024).
5. Slon, C.; Pandey, V. Enabling autonomous decision-making in manufacturing systems through preference fusion. *SAE Int. J. Mater. Manuf.* **2020**, *13*, 109–124. [CrossRef]
6. Monostori, L.; Kádár, B.; Bauernhansl, T.; Kondoh, S.; Kumara, S.; Reinhart, G.; Sauer, O.; Schuh, G.; Sihn, W.; Ueda, K. Cyber-physical systems in manufacturing. *Cirp Ann.* **2016**, *65*, 621–641. [CrossRef]
7. Schmitt, R.; Permin, E.; Kerkhoff, J.; Plutz, M.; Böckmann, M.G. Enhancing resiliency in production facilities through cyber physical systems. *Ind. Internet Things Cybermanuf. Syst.* **2017**, 287–313. [CrossRef]
8. Khan, U.; Haleem, A. Smart organisations: Modelling of enablers using an integrated ISM and fuzzy-MICMAC approach. *Int. J. Intell. Enterp.* **2012**, *1*, 248–269. [CrossRef]
9. Presciuttini, A.; Portioli-Staudacher, A. Applications of IoT and Advanced Analytics for manufacturing operations: A systematic literature review. *Procedia Comput. Sci.* **2024**, *232*, 327–336. [CrossRef]
10. Khan, U.; Haleem, A. Improving to smart organization: An integrated ISM and fuzzy-MICMAC modelling of barriers. *J. Manuf. Technol. Manag.* **2015**, *26*, 807–829. [CrossRef]
11. Canziani, A.; Paszke, A.; Culurciello, E. An analysis of deep neural network models for practical applications. *arXiv* **2016**, arXiv:1605.07678. [CrossRef]
12. Liu, C.; Jiang, P.; Jiang, W. Web-based digital twin modeling and remote control of cyber-physical production systems. *Robot. Comput.-Integr. Manuf.* **2020**, *64*, 101956. [CrossRef]

13. Qamsane, Y.; Chen, C.Y.; Balta, E.C.; Kao, B.-C.; Mohan, S.; Moyne, J.; Tilbury, D.; Barton, K. A Unified Digital Twin Framework for Real-Time Monitoring and Evaluation of Smart Manufacturing Systems. In Proceedings of the 2019 IEEE 15th International Conference on Automation Science and Engineering (CASE), IEEE, Vancouver, BC, Canada, 22–26 August 2019; pp. 1394–1401 [CrossRef]
14. Kessels, B.M.; Fey, R.H.; van de Wouw, N. Real-time parameter updating for nonlinear digital twins using inverse mapping models and transient-based features. *Nonlinear Dyn.* **2023**, *111*, 10255–10285. [CrossRef]
15. Rojek, I.; Jasiulewicz-Kaczmarek, M.; Piechowski, M.; Mikołajewski, D. An artificial intelligence approach for improving maintenance to supervise machine failures and support their repair. *Appl. Sci.* **2023**, *13*, 4971. [CrossRef]
16. Blume, C.; Blume, S.; Thiede, S.; Herrmann, C. Data-driven digital twins for technical building services operation in factories: A cooling tower case study. *J. Manuf. Mater. Process.* **2020**, *4*, 97. [CrossRef]
17. Liu, D.; Du, Y.; Chai, W.; Lu, C.; Cong, M. Digital twin and data-driven quality prediction of complex die-casting manufacturing *IEEE Trans. Ind. Inform.* **2022**, *18*, 8119–8128. [CrossRef]
18. Akinsolu, M.O.; Zribi, K. A generalized framework for adopting regression-based predictive modeling in manufacturing environments. *Inventions* **2023**, *8*, 32. [CrossRef]
19. Xu, Y.; Qamsane, Y.; Puchala, S.; Januszczak, A.; Tilbury, D.M.; Barton, K. A data-driven approach toward a machine-and system-level performance monitoring digital twin for production lines. *Comput. Ind.* **2024**, *157*, 104086. [CrossRef]
20. Rathore, M.M.; Shah, S.A.; Shukla, D.; Bentafat, E.; Bakiras, S. The role of ai, machine learning, and big data in digital twinning: A systematic literature review, challenges, and opportunities. *IEEE Access* **2021**, *9*, 32030–32052. [CrossRef]
21. Wang, P.; Luo, M. A digital twin-based big data virtual and real fusion learning reference framework supported by industrial internet towards smart manufacturing. *J. Manuf. Syst.* **2021**, *58*, 16–32. [CrossRef]
22. Aivaliotis, P.; Georgoulias, K.; Chryssolouris, G. The use of Digital Twin for predictive maintenance in manufacturing. *Int. J. Comput. Integr. Manuf.* **2019**, *32*, 1067–1080. [CrossRef]
23. Ho, S.L.; Xie, M.; Goh, T.N. A comparative study of neural network and Box-Jenkins ARIMA modeling in time series prediction. *Comput. Ind. Eng.* **2002**, *42*, 371–375. [CrossRef]
24. Ran, Y.; Zhou, X.; Lin, P.; Wen, Y.; Deng, R. A Survey of Predictive Maintenance: Systems, Purposes and Approaches. *arXiv* **2019**, arXiv:1912.07383. [CrossRef]
25. Papacharalampopoulos, A.; Michail, C.K.; Stavropoulos, P. Manufacturing resilience and agility through processes digital twin: Design and testing applied in the LPBF case. *Procedia Cirp* **2021**, *103*, 164–169. [CrossRef]
26. Khazraei, K.; Deuse, J. A strategic standpoint on maintenance taxonomy. *J. Facil. Manag.* **2011**, *9*, 96–113. [CrossRef]
27. Yu, W.; Dillon, T.; Mostafa, F.; Rahayu, W.; Liu, Y. A global manufacturing big data ecosystem for fault detection in predictive maintenance. *IEEE Trans. Ind. Inform.* **2019**, *16*, 183–192. [CrossRef]
28. Sakib, N.; Wuest, T. Challenges and opportunities of condition-based predictive maintenance: A review. *Procedia Cirp* **2018**, *78*, 267–272. [CrossRef]
29. He, Y.; Gu, C.; Chen, Z.; Han, X. Integrated predictive maintenance strategy for manufacturing systems by combining quality control and mission reliability analysis. *Int. J. Prod. Res.* **2017**, *55*, 5841–5862. [CrossRef]
30. Hosseinzadeh, A.; Chen, F.F.; Shahin, M.; Bouzary, H. A predictive maintenance approach in manufacturing systems via AI-based early failure detection. *Manuf. Lett.* **2023**, *35*, 1179–1186. [CrossRef]
31. Garcia, M.C.; Sanz-Bobi, M.A.; Del Pico, J. SIMAP: Intelligent System for Predictive Maintenance: Application to the health condition monitoring of a windturbine gearbox. *Comput. Ind.* **2006**, *57*, 552–568. [CrossRef]
32. Akhavei Bleicher, F. Predictive Modeling to Increase the Reliability of Production Planning in Single-Item Production. In Proceedings of the World Congress on Engineering and Computer Science 2016, San Francisco, CA, USA, 19–21 October 2016; Volume II, pp. 806–811.
33. Lee, J.; Davari, H.; Singh, J.; Pandhare, V. Industrial Artificial Intelligence for industry 4.0-based manufacturing systems. *Manuf. Lett.* **2018**, *18*, 20–23. [CrossRef]
34. Rahman, M.S.; Ghosh, T.; Aurna, N.F.; Kaiser, M.S.; Anannya, M.; Hosen, A.S. Machine learning and internet of things in industry 4.0: A review. *Meas. Sens.* **2023**, *28*, 100822. [CrossRef]
35. De Simone, V.; Di Pasquale, V.; Miranda, S. An overview on the use of AI/ML in manufacturing MSMEs: Solved issues, limits, and challenges. *Procedia Comput. Sci.* **2023**, *217*, 1820–1829. [CrossRef]
36. Canal, A.D.; Borille, A.V. CNC Turning: Roughness, Forces and Tool Wear [Dataset]. 2021 [Cited 2022]. In Kaggle [Internet]. [CrossRef]
37. Plevris, V.; Solorzano, G.; Bakas, N.P.; Ben Seghier, M.E. Investigation of Performance Metrics in Regression Analysis and Machine Learning-Based Prediction Models. In Proceedings of the 8th European Congress on Computational Methods in Applied Sciences and Engineering (ECCOMAS Congress 2022), Oslo, Norway, 5–9 June 2022. [CrossRef]
38. Mienye, I.D.; Sun, Y. A survey of ensemble learning: Concepts, algorithms, applications, and prospects. *IEEE Access* **2022**, *10*, 99129–99149. [CrossRef]

39. Chicco, D.; Warrens, M.J.; Jurman, G. The coefficient of determination R-squared is more informative than SMAPE, MAE, MAPE, MSE and RMSE in regression analysis evaluation. *Peerj Comput. Sci.* **2021**, *7*, e623. [CrossRef]
40. Willmott, C.J.; Matsuura, K. Advantages of the mean absolute error (MAE) over the root mean square error (RMSE) in assessing average model performance. *Clim. Res.* **2005**, *30*, 79–82. [CrossRef]
41. Chai, T.; Draxler, R.R. Root mean square error (RMSE) or mean absolute error (MAE)?—Arguments against avoiding RMSE in the literature. *Geosci. Model Dev.* **2014**, *7*, 1247–1250. [CrossRef]
42. Kim, S.; Kim, H. A new metric of absolute percentage error for intermittent demand forecasts. *Int. J. Forecast.* **2016**, *32*, 669–679. [CrossRef]
43. Das, B.R.; Resnick, S.I. QQ plots, random sets and data from a heavy tailed distribution. *Stoch. Models* **2008**, *24*, 103–132. [CrossRef]

machines

Article

Obstacle Detection for Autonomous Guided Vehicles through Point Cloud Clustering Using Depth Data

Micael Pires [1], Pedro Couto [1,2,*], António Santos [3] and Vítor Filipe [1,4]

1 School of Science and Technology, University of Trás-os-Montes e Alto Douro (UTAD), 5000-801 Vila Real, Portugal; piresmicael@gmail.com (M.P.); vfilipe@utad.pt (V.F.)
2 CITAB—Centre for the Research and Technology of Agro-Environmental and Biological Sciences, Quinta de Prados, 5000-801 Vila Real, Portugal
3 Active Space Technologies, Parque Industrial de Taveiro, Lote 12, 3045-508 Coimbra, Portugal; antonio.santos@activespacetech.com
4 Institute for Systems and Computer Engineering Technology and Science (INESC TEC), 4200-465 Porto, Portugal
* Correspondence: pcouto@utad.pt

Abstract: Autonomous driving is one of the fastest developing fields of robotics. With the ever-growing interest in autonomous driving, the ability to provide robots with both efficient and safe navigation capabilities is of paramount significance. With the continuous development of automation technology, higher levels of autonomous driving can be achieved with vision-based methodologies. Moreover, materials handling in industrial assembly lines can be performed efficiently using automated guided vehicles (AGVs). However, the visual perception of industrial environments is complex due to the existence of many obstacles in pre-defined routes. With the INDTECH 4.0 project, we aim to develop an autonomous navigation system, allowing the AGV to detect and avoid obstacles based in the processing of depth data acquired with a frontal depth camera mounted on the AGV. Applying the RANSAC (random sample consensus) and Euclidean clustering algorithms to the 3D point clouds captured by the camera, we can isolate obstacles from the ground plane and separate them into clusters. The clusters give information about the location of obstacles with respect to the AGV position. In experiments conducted outdoors and indoors, the results revealed that the method is very effective, returning high percentages of detection for most tests.

Keywords: obstacle detection; RANSAC algorithm; Euclidean clustering; 3D point cloud; RGB-D data

1. Introduction

Automation is one of the fastest-developing fields of robotics. With the ever-growing interest in automating processes, concerning both efficiency and safety, the ability to provide robots with autonomous navigation capabilities is of paramount significance [1,2].

Several sensing technologies, types of navigation systems and data fusion are available for the autonomous navigation of mobile robots. A key aspect of mobile autonomous robots is the ability to detect objects in their path and allow them to navigate freely with minimum risk of collision. Thus, in industrial environments, AGVs are usually provided with laser scanners to prevent them from running into people or objects. However, the effectiveness of these systems outdoors can be affected by adverse weather conditions such as direct sunlight, rain, snow or fog. Even considering the use of laser scanners prepared to operate outdoors, other vehicles (e.g., forklifts, truck trailers) may not be detected, possibly leading to collisions. To mitigate this problem, 3D point clouds captured with a depth camera could be used in these complex conditions. With 3D point clouds, more than ten thousand points of information, including 3D coordinates and reflection intensities, can be obtained in a short streaming sequence. Unlike RGB images, the significant advantages of point clouds are the obvious 3D spatial structures that also include reflection features about the surface of objects [3]. A large number of automated reconstruction approaches focus on the unsupervised processing of point clouds. The interpretation of these data is challenging due to the number of points, noise and the complexity of the structure [4]. Moreover, most point clouds are acquired with remote sensing techniques associated to line of sight (LoS). As a result, crucial parts of the structure are occluded due to clutter or inaccessible areas [5].

There are different ways to perform object detection using 3D point clouds. Most solutions tackle this problem with neural networks adapted to this kind of data. Several methods, depending on the type of deep learning network inputs that can be raw point cloud data [6], voxel [7] or image [8], have been proposed for 3D object detection. Shi et al. [6] proposed PointRCNN for 3D object detection from raw point clouds. The whole framework is composed of two stages: stage one for the bottom-up 3D proposal generation and stage two for refining proposals in the canonical coordinates to obtain the final detection results. Those results showed that the proposed architecture outperforms state-of-the-art methods with remarkable margins by using only the point cloud as input. An important study in this type of method by Zhou et al. presented the VoxelNet algorithm [9], which works with sparse 3D point clouds without the need of a manual feature selection. Further studies have led to the development of other algorithms (e.g., Qi et al. [10]) that use RGB images to generate detection proposals for 3D detection. However, ConvNets-based models generally require a large amount of data for effective training.

Ground plane extraction and obstacle detection are essential tasks for collision-free navigation of autonomous mobile robots [11]. One of the most common methods to detect the ground plane in a scene is based in processing depth information. To address the ground plane detection, Choy et al. [12] proposed a robust and fast ground plane detection with an asymmetric kernel and RANSAC (random sample consensus). A probabilistic model of 3D point clouds with the assumption that a point from other objects is always above the ground was derived. The asymmetric kernel is the key approximation for fast computation, which incorporates RANSAC as a score function. Experimental results showed that the method was sufficiently accurate.

After ground plane extraction, objects can be detected in three-dimensional data by using point cloud segmentation, which is the method explored in this work. Segmentation is one of the most important steps for the automatic processing of point clouds. It is a process of classifying and labelling data points into a set of separate groups or regions, each one ideally corresponding to an object. An example of this kind of methodology is discussed by Habermann et al. in [13], where they presented three different point cloud segmentation methods and compared their performance with the k-means segmentation algorithm. Miao et al. proposed a novel point-cloud-based 3D object detection method for achieving higher-accuracy of autonomous driving in [3]. The proposed method employed YOLOv4 as the feature extractor and Normal-map as additional input. This way, the additional input delivered more enhanced shape information and can be associated with other hand-crafted features, achieving higher precision for 3D object detection with lower perturbations from other parameters such as distance. Bizjak et al. proposed a new method for the segmentation of point clouds in [14]. The topological structuring of the input point cloud was realized by a nearest-neighbor graph, while locally fitted surfaces were used to quantify the objects' shapes. Noise was filtered by a graph cut, and the segmentation was performed by linking the nodes to those objects to which they fit best. The presented results showed that the proposed method can improve the accuracy of the segmentation while reducing the computation time in comparison with other methods.

A ground-aware framework that aims to solve the ambiguity caused by data sparsity present in lidar point cloud measurements was proposed by In Wu et al. in [15]. A multi-section plane fitting approach to roughly extract ground points to assist segmentation of objects on the ground was employed. The proposed method captures long-range dependence between the ground and objects, which significantly facilitates the segmentation of small objects that only consist of a few points in extremely sparse point clouds.

Results showed that the proposed method outperforms state-of-the art methods such as PointNet [16] and SPG [17].

Lu et al. [18] proposed a method that proved to be a robust and efficient point cloud segmentation algorithm in a large set of experimental results. A novel hierarchical clustering algorithm named Pairwise Linkage (P-Linkage) was proposed. The method can be used for clustering any dimensional data and then effectively applied on 3D unstructured point cloud segmentation.

From a different perspective, based on the argument that a single segmentation method will typically not provide a satisfactory segmentation for a variety of classes, Vosselman et al. [19] proposed a methodology that explores the combination of various segmentation and post-processing methods to arrive at useful point cloud segmentations. They concluded that by combining different segmentation and post-processing methods, a segmentation result can be obtained for group points of different classes (e.g., vegetation, ground, walls, roofs), hence increasing the quality of a segment-based classification. A 3D point cloud segmentation survey was presented by Nguyen et al. in [20], where a comparative analysis on various state-of-the-art point cloud segmentation methods is made. Furthermore, Liang et al. presented a survey where several 3D object detection methods based on point cloud are described and analyzed [21]; in addition, those methods were compared in terms of advantages and disadvantages for 3D object detection, including image-based and fusion of point cloud and image methods.

In the project INDTECH 4.0, developed by a consortium of companies and universities, one of the work packages aims to study and develop autonomous navigation solutions to be integrated in AGVs operating in the automotive assembly industry. In this context, an Intel RealSense D435i camera was mounted on the front of an AGV to collect depth data. The 3D point clouds acquired are continuously processed to ensure obstacle avoidance, which is an essential requirement for real-time outdoor autonomous navigation of AGVs.

In this paper, we present a method to detect potential obstacles and to measure their relative position with respect to the robot, using point cloud data provided by the depth camera [22]. This way, the AGV is guided through vision technology, without any previous knowledge of the environment that can be static or dynamic, indoor or outdoor, in a process called natural navigation.

The remainder of this paper is structured as follows. Section 2 presents a description of our methodology, detailing the various processing phases of our method. Section 3 discusses the results according to various metrics relevant for this kind of algorithm. Section 4 summarizes the conclusions and suggestions for future work.

2. Proposed Methodology

In this work, we aim to develop a method for obstacle detection in front of an AGV and measure their positions to the camera. To that end, we used a depth camera, which captures point clouds representing the surrounding space. Since these point clouds contain the relevant features of the robot's surrounding environment regarding potential obstacles, it is important to pose the camera in such a way that obstacles in the path of the AGV can be detected. It is concluded that the most interesting placement for the camera is at the front of the AGV, so that we capture the full extent of the obstacles in front and reduce the number of points corresponding to the ground plane. During the tests we also empirically found the optimal height and pitch for the camera. The experimental setup can be seen in Figure 1.

As the point clouds contain all elements of the scene, some processing has to be carried out in order to remove scene elements that are not interesting for obstacle detection, namely the ground plane. By extracting the ground plane, the output point cloud is made of a more detached set of clusters of points that represent various objects in the scene. These sets of points can be grouped through clustering algorithms into groups that we use as evidence to an existent obstacle and for position measuring. To operate in real time, the point clouds recorded with the camera were subsequently range limited and down sampled to reduce computational effort.

A method to detect obstacles and their locations by means of 3D point clouds provided by the Intel RealSense D435i depth camera is proposed. This point cloud is subjected to three different stages of processing.

Figure 1. Experimental setup used during this work.

2.1. Pass-Through Filter and Voxel Grid

The first processing stage consists of applying a pass-through filter. This filter is used to restrict the point cloud limits to a relevant distance window.

The ultimate use of this algorithm is to apply it in a navigation system of a low-speed AGV for logistics. Thus, it is possible to limit the computational effort while conserving a high relevance of the results since, due to the low speed, we do not need to detect obstacles very far from the robot.

Another processing tool used to save computational time is the voxel grid to down sample the input point cloud. This filter divides the point cloud into voxels with a user-defined size. The points present in each voxel are approximated by their centroid. This process results in a point cloud with fewer points that still describes the original surface accurately.

In our tests, we defined the limits of the distance window in the interval [1.5, 5] m along the z-axis (depth) for the pass-through filter and a voxel edge length of 0.07 m for the voxel grid.

2.2. RANSAC Segmentation

To successfully identify obstacles in the camera's field of view, it is necessary to segment the volume of the point cloud that corresponds to the ground plane. This is carried out to deliver a higher degree of separation between objects, facilitating the process of grouping points into clusters. For this purpose, the RANSAC (random sample consensus) algorithm is used [23].

RANSAC is an iterative algorithm for fitting mathematical models to experimental data. Instead of considering all the data and eliminating invalid data points, as common in most conventional smoothing techniques, RANSAC uses the minimum amount of data possible and adds as consistent an amount of data points as possible [11]. This method takes the minimum number of points that define a particular model from the experimental set and instantiates a model using those points, (in the case of plane detection, RANSAC selects three random points from the set). Having a candidate model defined, the number of points that are within a defined tolerance to the plane is counted (inliers). When the number of inliers exceeds a threshold, the candidate model is chosen. If the number of inliers is too low, the process is repeated until a plane is identified or the maximum number of iterations is achieved.

The PCL (point cloud library) implementation of RANSAC further includes the option of constraining the plane fit to a specified axis within an angular tolerance. This helps in identifying the ground plane by restricting the detection to horizontal planes. We set the angular tolerance as six degrees.

Once the ground plane is identified, inliers are removed, and the resulting point cloud solely contains objects that may represent obstacles.

2.3. Euclidean Clustering

After removing the ground plane, Euclidean clustering is carried out to identify the various clusters of points present in the point cloud. This is a crucial step to discard points that do not belong to a meaningful cluster and thus do not represent a potential obstacle.

Euclidean clustering consists of separating sets of points according to the Euclidean distance between those points. If two points are within a predefined distance threshold, they are considered to be in the same cluster. However, if that distance is higher than the threshold, the points are segregated into different clusters. Mathematically, we can describe the problem of finding clusters as follows. Letting P be our input dataset and d_{th} a defined distance threshold, a particular set of points $O_i = \{p_i \in \mathrm{P}\}$ is a distinct cluster to $O_j = \{p_j \in \mathrm{P}\}$ if $min||p_i - p_j||_2 \geq d_{th}$ [24].

The PCL implementation of Euclidean clustering also contains the ability to define the minimum and maximum cluster sizes to mitigate the detection of undesirable sets of points.

Once all the clusters are defined, we can visualize them and measure the position to each obstacle. For the interest of avoiding AGV collisions, our method returns the coordinates of the closest point of each cluster.

2.4. Implementation

The method processing sequence is shown in Figure 2.

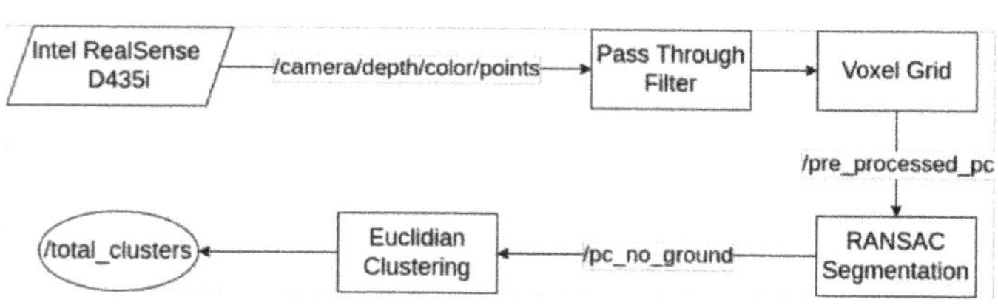

Figure 2. Processing flowchart.

The method consists of an ROS (robot operating system) node that subscribes to the point cloud provided directly by the depth camera in the "camera/depth/color/points" ROS topic. Some intermediate topics are also published for algorithm visualization. These topics include the "pre_processed_pc" topic for the down-sampled and range-limited point cloud and the "pc_no_ground" topic for the point cloud after removal of the ground plane.

The result is the separation of obstacles into point clusters. Each cluster represents a different obstacle, and through the clusters we can obtain the position of each obstacle relative to the AGV. After processing, all detected clusters are merged in a point cloud that is published in the "total_clusters" ROS topic. An additional ROS topic is published for analysis of the algorithm's performance: the "cluster_closest_point" topic that is discussed in the results section.

In addition to ROS, functionalities provided by PCL [25] have been used in all processing steps. These functionalities include the down sampling, plane detection and extraction and clustering algorithms.

3. Results

3.1. Experimental Data

Throughout this work, various tests have been carried out to optimize this algorithm. These tests have been performed on the premises of Active Space Technologies and consisted of recording ROS bags using the D435i depth camera with various types of obstacles located at different distances. We tested the algorithm for both static and mobile obstacles considering indoor and outdoor environments.

To represent static obstacles, chairs placed at 1.8 m, 2.5 m and 4.5 m were used. The distance to the chairs was measured taking the closest part of the chair into account. This setup allowed us to evaluate the algorithm performance for different distances across the defined depth window ([1.5, 5] m). For each of the three distances, an ROS bag was recorded with two chairs side-by-side, indoors and outdoors.

Another test was made with three chairs, one at each distance mentioned above, to assess simultaneous detection of various obstacles at different distances. The final test with chairs consisted of moving the camera perpendicularly to the three chairs, in the direction of the chair furthest away. This test simulates object detection of static obstacles with a moving AGV.

For mobile obstacles, we performed tests in which people moved in front of the camera in various directions. A final test was made where the camera was also moving among wandering people, simulating a more complex scenario that would be common for an industrial AGV.

Every test mentioned above was performed indoors and repeated outdoors. This batch of tests enables us to obtain a sense of the algorithm's performance in different conditions considered relevant for the project. Figures 3 and 4 represent the typical outputs of this algorithm concerning the three major processing stages.

Figure 3. Indoor results for the four stages of processing. From top-left to bottom-right: original point cloud, down-sampled and range-limited point cloud, point cloud after ground plane removal and point cloud with detected clusters (two chairs).

Figure 4. Outdoor results for the four stages of processing.

3.2. Object Detection Effectiveness

There are various 3D object detection frameworks, each one of them using different imaging technologies and methods and operating under different conditions. For these reasons, a direct comparison between the proposed framework and the existing ones could lead to misleading conclusions. To correctly analyze the performance of different frameworks, the same datasets (both indoor and outdoor datasets) under different metrics should be used [21].

To measure the object detection effectiveness, we counted the number of times that each object was detected and compared it with the total number of frames of each test. This was performed using the ROS bags corresponding to the placement of chairs side-by-side at various distances and using indoor and outdoor measurements.

For each ROS bag, the point from each cluster closest to the camera was obtained and published in an ROS topic (cluster_closest_point) for analysis. This topic contains the xyz coordinates of the closest point of each object.

The total number of point clouds analyzed by the algorithm was also counted.

A Python script was used to subscribe to the (cluster_closest_point) topic, and the information was assembled and plotted in a 3D graph. This graph gave us a sense of the position of obstacles in the corresponding test.

In addition to visualizing the data, a clustering algorithm was applied to this set of points, originating clusters of points that correspond to detections of the same object. Using the clusters of closest points, we counted the number of points that represent each obstacle detected by the camera. That value determines how many times each obstacle is detected.

By comparing these values with the number of point clouds processed by our object detection algorithm, we were able to calculate the effectiveness of the object detection.

Each ROS bag has a duration of approximately 15 s, which at $\simeq$30 fps represents a total of 450 frames for each test run. The process described above was repeated 10 times for each ROS bag, and the results are presented in Tables 1 and 2.

Table 1. Indoor detection effectiveness.

Distance (m)	Indoor Detection Effectiveness (%)
1.8	92.72
2.5	70.59
4.5	51.02

Table 2. Outdoor detection effectiveness.

Distance (m)	Outdoor Detection Effectiveness (%)
1.8	99.05
2.5	99.12
4.5	89.22

From these results, we concluded that the method is very effective, returning high percentages of detection for most tests. Moreover, the percentages of detection of both objects were very similar for all distances.

It can be seen that the detection rates drop more significantly for indoor obstacles farther away from the camera. This is due to the poorer luminosity and the fact that for farther objects, the point cloud includes parts of the ceiling, extending the dimensions of the point cloud and making the object detection process more difficult. In this case, the performance bottleneck arises from the increased size of the point cloud, hence explaining why, for outdoor conditions, an efficacy drop is not as drastic.

We can also conclude that, albeit the detection rates varied according to environmental conditions, there was not a variation in detection efficacy between the two obstacles. Thus, detection success decreases with distance to the obstacle. The percentage of obstacle detection is also significantly more stable in outdoor conditions.

3.3. Distance Measurements

A major requirement of this work was not only the detection of obstacles in front of the robot but also measuring their position relative to the camera.

Using the same ROS bags, with two chairs side-by-side at various distances, we analyzed the distance measurements obtained with our algorithm.

The results in Tables 3 and 4 show the mean and standard deviation for the distances obtained in each test, for indoor and outdoor conditions.

Table 3. Indoor detection effectiveness.

Indoor Measurements		
Distance (m)	**Average Distance (m)**	**Standard Deviation (m)**
1.8	1.85	0.03
2.5	2.60	0.09
4.5	5.01	0.11

Table 4. Outdoor detection effectiveness.

Outdoor Measurements		
Distance (m)	**Average Distance (m)**	**Standard Deviation (m)**
1.8	1.91	0.03
2.5	2.56	0.13
4.5	4.95	0.20

For distance measurements, we were able to obtain reasonable values, with an expected drop in accuracy for farther objects. These measurements are entirely dependent on the accuracy of the Intel RealSense D435i depth camera, and as demonstrated, we can rely on the good accuracy provided by the depth sensor for our application.

3.4. Moving Object Detection

For moving obstacles, the performance of the algorithm shows favorable results, although the analysis of these tests was made merely by visual inspection.

The proposed method is able to segment and follow moving obstacles smoothly with a high degree of accuracy and with a high frame rate without apparent failure in detection.

Figure 5 shows two people walking side-by-side in opposite directions in front of the stationary AGV; both people are detected and segmented correctly.

Figure 5. Moving obstacle detection (pedestrians walking in opposite directions). In this test, the AGV was stationary. (**Top**): original point cloud; (**Bottom**): detected clusters.

Figure 6 shows two frames from the mobile AGV/wandering person test where it is clear that the person is segmented accurately. In most frames, there are no false positives. The detection remains accurate during the whole test regardless of the movement of the AGV and of the person.

Figure 6. Moving obstacle detection (wandering pedestrian). In this test, both the person and the AGV were moving. (**Top**): original point clouds; (**Bottom**): detected clusters.

3.5. Frame Rates

Since this algorithm is intended for real-time obstacle avoidance, it was also important to analyze the frame rate.

Using a computer with an Intel® Core™ i7-6700HQ CPU @ 2.60 GHz, we did not register significant frame rate drops relative to the native frame rate of the point cloud ($\simeq$30 fps), with the maximum frame rate drop being of $\simeq$58% in one of the tests but that results were still in a rate of $\simeq$12 fps for obstacle detection. That being said, most of the tests performed had a frame drop inferior to 10% relative to the native frame rate of the camera, with the average frame drop in all tests being of 18%. These values mean that our method is applicable in a slow moving AGV for real-time object detection.

Table 5 shows data regarding the frame rates from all tests pictured above and two extra tests where we obtained the best and worst results regarding frame rate drops. The columns contain the frame rates for the ROS topics of the original point cloud provided by the camera ("camera/depth/color/points") and of the final topic containing all clusters ("total_clusters").

Table 5. Frame rate drops for different tests.

	Frame Rates		
Test	**Point Cloud**	**Clusters**	**Frame Rate Drop (%)**
Indoor chair detection	30.05	27.12	9.77
Outdoor chair detection	18.37	17.71	3.56
Moving people/stationary AGV	29.94	22.20	25.84
Moving person/moving AGV	29.98	27.34	8.83
Highest frame rate drop	29.96	12.32	58.88
Lowest frame rate drop	30.02	29.40	2.09

4. Discussion

In this work, an object detection algorithm using point cloud data from an Intel® RealSense™ D435i is proposed. This algorithm is meant to be applied in an autonomous navigation system with obstacle avoidance capabilities for AGVs working in dynamic industrial environments.

The presented results show this method can detect obstacles and measure their position relative to the AGV in various environmental conditions, from low luminosity to sunny conditions. This is very important in a SLAM (simultaneous localization and mapping) context to enable the robust autonomous navigation of robots. This method is also able to function in real time, considering the pre-processing applied to the point clouds provided by the camera, which is crucial for autonomous navigation. In this context, real-time conditions mean the algorithm is capable of processing and making decisions quickly enough to ensure the AGV can be safely driven when it is moving with a velocity up to 1.5 m s^{-1}. The latency of the system, i.e., the time delay between the cause and the effect of some physical change (e.g., obstacle motion) being observed, is usually lower than 50 ms.

The algorithm detects objects with high reliability and returns results with inconsequential frame drops from the camera's native frame rate. When the illumination conditions are optimal, no frame rate drop has been observed, at least not to compromise system performance. This applies for the various environmental conditions tested, specifically indoor and outdoor environments, static and mobile obstacles and with the AGV being mobile or stationary. The resulting frame rates of detection are adequate for a slow-moving AGV in an industrial environment, as intended for our application. Supplementary tests have been made to ascertain the impact of direct illumination on the sensor. As long as direct incident radiation from the source (e.g., sunlight) does not hit the optical sensor directly, the algorithm is capable of reliable object detection. The exception has been a camera malfunction due to indirect heating effects rather than illumination issues of the sensor.

The obstacle position measurements returned by our method also present high degrees of precision that improve with illumination and when obstacles get closer to the camera. In some cases, camera pitch was altered to optimize the detection range, but the algorithm performance remained stable. This was shown with the outdoor tests, where the measurements presented lower standard deviation. These are characteristics innate to the Intel RealSense D435i depth camera, and the precision of distance measurements is entirely dependent on the capabilities of the chosen camera to be used along with our obstacle detection method.

Thus, a three-stage methodology using well-established methodologies such as a pass-through filter and voxel grid for data dimensionality reduction, RANSAC algorithm for identification and removal of inliers of the ground plane and Euclidean clustering for point cloud clustering for object detection is proposed. As a result, a framework for real-time autonomous navigation with low computational effort and no need for prior knowledge of the surrounding environment is proposed in this work.

In addition to indoor measurements frequently reported in many studies that consider stable, calibrated illumination conditions, we have conducted a test campaign considering several outdoor experiments to assess algorithm performance and identify possible limitations and blocking points. Although carried out at a time when the factory was not under nominal operation conditions, logistics and environmental concerns were selected to mimic real environments. The most relevant conditions were the following: ram- and wake-facing natural illumination, clear and cloudy sky, flat and inclined paths (slope up to 8%), smooth and irregular grounds, different types of objects and vehicles as well as multiple pedestrians wandering in different directions. The algorithm was able to perform correctly under all these conditions without major issues. Qualitatively, algorithm performance was not degraded during the whole campaign, at least not to a point rendering it useless. As mentioned above, the main issue was caused by a temporary hardware failure (camera overheating) rather than poor performance of the algorithm.

Further developments of this algorithm may provide the capability to create a system that generates odometry measurements using depth camera data, namely by implementing a visual odometry system for SLAM. Visual odometry systems calculate the movement of a robot by taking advantage of information regarding obstacle position over time, which we can obtain through our method. Such algorithms are useful to perform sensor fusion with an inertial measurement unit or, according to the accuracy of their results, can even replace this kind of sensors.

5. Conclusions

Having achieved a high reliability regarding object detection, the presented methodology can be extended to process RGB images, acquired simultaneously by the Intel camera, to extend the capabilities of our algorithm to not only detect the position of obstacles but also classify them into types of objects. For indoor detection, being a controlled environment, previous knowledge about possible/existing obstacles (e.g., person, forklift, bucket, etc.) can be used to eliminate potential obstacles and improve effectiveness, being that the RGB images can be easily used to classify objects using that prior knowledge. With the information of the type of obstacle in front of the robot, it is possible to program more informed decisions regarding the movement of the AGV. Another interesting topic for further development is the detection of holes in the ground, which is another crucial problem that must be solved for safe AGV movement, and it is a feature that is usually not detected with traditional laser scanners.

Author Contributions: Conceptualization, M.P., A.S., P.C. and V.F.; data curation, M.P.; investigation, M.P.; methodology, M.P., P.C., A.S. and V.F.; supervision, V.F.; validation, M.P., P.C., A.S. and V.F.; writing—original draft preparation, M.P.; review and editing, M.P., P.C., A.S. and V.F. All authors have read and agreed to the published version of the manuscript.

Funding: This work was funded by Project "INDTECH 4.0—new technologies for smart manufacturing", POCI-01-0247-FEDER-026653, and by the European Regional Development Fund (ERDF) through the COMPETE 2020—Competitiveness and Internationalization Operational Program (POCI). This work was also funded by national funds from FCT—Portuguese Foundation for Science and Technology, under the project UIDB/04033/2020.

Informed Consent Statement: Informed consent was obtained from all subjects involved in the study.

Data Availability Statement: Not applicable.

Conflicts of Interest: The authors declare no conflict of interest. The funders had no role in the design of the study; in the collection, analyses or interpretation of data; in the writing of the manuscript or in the decision to publish the results.

References

1. Pires, M.A. Natural Navigation Solutions for AMRs and AGVs Using Depth Cameras. Master's Thesis, University of Coimbra, Coimbra, Portugal, 2021.
2. Lynch, L.; Newe, T.; Clifford, J.; Coleman, J.; Walsh, J.; Toal, D. Automated ground vehicle (AGV) and sensor technologies—A review. In Proceedings of the 2018 12th International Conference on Sensing Technology (ICST), Limerick, Ireland, 4–6 December 2018; pp. 347–352.
3. Miao, J.; Hirakawa, T.; Yamashita, T.; Fujiyoshi, H. 3D Object Detection with Normal-map on Point Clouds. In Proceedings of the 16th International Joint Conference on Computer Vision, Imaging and Computer Graphics Theory and Applications, online, 8–10 February 2021; pp. 569–576. [CrossRef]
4. Tang, P.; Huber, D.; Akinci, B.; Lipman, R.; Lytle, A. Automatic reconstruction of as-built building information models from laser-scanned point clouds: A review of related techniques. *Autom. Constr.* **2010**, *19*, 829–843. [CrossRef]
5. Bassier, M.; Vergauwen, M. Clustering of Wall Geometry from Unstructured Point Clouds. In Proceedings of the International Archives of the Photogrammetry, Remote Sensing and Spatial Information Sciences, Bergamo, Italy, 6–10 May 2019; Volume XLII-2/W9, pp. 101–108. [CrossRef]
6. Shi, S.; Wang, X.; Li, H. PointRCNN: 3d object proposal generation and detection from point cloud. In Proceedings of the IEEE Conference on Computer Vision and Pattern Recognition, Long Beach, CA, USA, 15–20 June 2019; pp. 770–779.
7. Simon, M.; Amende, K.; Kraus, A.; Honer, J.; Samann, T.; Kaulbersch, H.; Milz, S.; Michael Gross, H. Complexer-yolo: Real-time 3d object detection and tracking on semantic point clouds. In Proceedings of the IEEE Conference on Computer Vision and Pattern Recognition Workshops, Long Beach, CA, USA, 15–20 June 2019.
8. Beltrán, J.; Guindel, C.; Moreno, F.M.; Cruzado, D.; García, F.; Escalera, A.D.L. Birdnet: A 3d object detection framework from lidar information. In Proceedings of the 2018 21st International Conference on Intelligent Transportation Systems (ITSC), Maui, HI, USA, 4–7 November 2018; pp. 3517–3523.

9. Zhou, Y.; Tuzel, O. Voxelnet: End-to-end learning for point cloud based 3D object detection. In Proceedings of the IEEE Conference on Computer Vision and Pattern Recognition, Salt Lake City, UT, USA, 18–22 June 2018; pp. 4490–4499.
10. Qi, C.R.; Liu, W.; Wu, C.; Su, H.; Guibas, L.J. Frustum pointnets for 3D object detection from RGB-D data. In Proceedings of the IEEE Conference on Computer Vision and Pattern Recognition, Salt Lake City, UT, USA, 18–22 June 2018; pp. 918–927.
11. Kırcalı, D.; Tek, F.B. Ground plane detection using an RGB-D sensor. In *Information Sciences and Systems*; Springer: Cham, Switzerland, 2014; pp. 69–77.
12. Choi, S.; Park, J.; Byun, J.; Yu, W. Robust Ground Plane Detection from 3D Point Clouds. In Proceedings of the 2014 14th International Conference on Control, Automation and Systems (ICCAS 2014), Seoul, Korea, 22–25 October 2014; pp. 1076–1081.
13. Habermann, D.; Hata, A.; Wolf, D.; Osório, F.S. 3D point clouds segmentation for autonomous ground vehicle. In Proceedings of the 2013 III Brazilian Symposium on Computing Systems Engineering, Niteroi, Brazil, 4–8 December 2013; pp. 143–148.
14. Bizjak, M. The segmentation of a point cloud using locally fitted surfaces. In Proceedings of the 18th Mediterranean Electrotechnical Conference, Lemesos, Cyprus, 18–20 April 2016.
15. Wu, J.; Jiao, J.; Yang, Q.; Zha, Z.; Chen, X. Ground-Aware Point Cloud Semantic Segmentation for Autonomous Driving. In Proceedings of the 27th ACM International Conference on Multimedia (MM'19), Nice, France, 21–25 October 2019; pp. 971–979. [CrossRef]
16. Charles, R.; Su, H.; Kaichun, M.; Guibas, L. PointNet: Deep Learning on Point Sets for 3D Classification and Segmentation. In Proceedings of the IEEE Conference Computer Vision and Pattern Recognition (CVPR), Honolulu, HI, USA, 21–26 July 2017; pp. 77–85.
17. Landrieu, L.; Simonovsky, M. Large-scale point cloud semantic segmentation with superpoint graphs. In Proceedings of the IEEE Conference Computer Vision and Pattern Recognition (CVPR), Salt Lake City, UT, USA, 18–23 June 2018; pp. 4558–4567.
18. Lu, X.; Yao j Tu, J.; Li, K.; Li, L.; Liu, Y. Pairwise Linkage for Point Cloud Segmentation. In Proceedings of the ISPRS Annals of the Photogrammetry, Remote Sensing and Spatial Information Sciences, Prague, Czech Republic, 12–19 July 2016; Volume III-3, pp. 201–208. [CrossRef]
19. Vosselman, G. Point cloud segmentation for urban scene classification. *Int. Arch. Photogramm. Remote Sens. Spat. Inf. Sci.* **2013**, *1*, 257–262. [CrossRef]
20. Nguyen, A.; Le, B. 3D point cloud segmentation: A survey. In Proceedings of the 2013 6th IEEE Conference on Robotics, Automation and Mechatronics (RAM), Manila, Philippines, 12–15 November 2013; pp. 225–230.
21. Liang, W.; Xu, P.; Guo, L.; Bai, H.; Zhou, Y.; Chen, F. A survey of 3D object detection. *Multimed. Tools Appl.* **2021**, *80*, 29617–29641. [CrossRef]
22. Pires, M.; Couto, P.; Santos, A.; Filipe, V. Obstacle Segmentation for Autonomous Guided Vehicles through Point Cloud Clustering with an RGB D Camera. In Proceedings of the 2nd IFSA Winter Conference on Automation, Robotics and Communications for Industry 4.0 (ARCI'2022), Andorra la Vella, Andorra, 2–4 February 2022; pp. 79–83.
23. Fischler, M.A.; Bolles, R.C. Random sample consensus: A paradigm for model fitting with applications to image analysis and automated cartography. *Commun. ACM* **1981**, *24*, 381–395. [CrossRef]
24. Rusu, R.B. Semantic 3D object maps for everyday manipulation in human living environments. *KI-Künstl. Intell.* **2010**, *24*, 345–348. [CrossRef]
25. Rusu, R.B.; Cousins, S. 3D is here: Point cloud library (PCL). In Proceedings of the 2011 IEEE International Conference on Robotics and Automation, Shanghai, China, 9–13 May 2011; pp. 1–4.

machines

MDPI

Article

Autonomous Visual Navigation for a Flower Pollination Drone

Dries Hulens [1,†], Wiebe Van Ranst [1,*,†], Ying Cao [2] and Toon Goedemé [1,*]

1 EAVISE (Embedded and Artificially intelligent VISion Engineering), KU Leuven, 2860 Sint-Katelijne-Waver, Belgium; dries.hulens@kuleuven.be
2 Magics Technologies NV, 2440 Geel, Belgium; ying.cao@magics.tech
* Correspondence: wiebe.vanranst@kuleuven.be (W.V.R.); toon.goedeme@kuleuven.be (T.G.)
† These authors contributed equally to this work.

Abstract: In this paper, we present the development of a visual navigation capability for a small drone enabling it to autonomously approach flowers. This is a very important step towards the development of a fully autonomous flower pollinating nanodrone. The drone we developed is totally autonomous and relies for its navigation on a small on-board color camera, complemented with one simple ToF distance sensor, to detect and approach the flower. The proposed solution uses a DJI Tello drone carrying a Maix Bit processing board capable of running all deep-learning-based image processing and navigation algorithms on-board. We developed a two-stage visual servoing algorithm that first uses a highly optimized object detection CNN to localize the flowers and fly towards it. The second phase, approaching the flower, is implemented by a direct visual steering CNN. This enables the drone to detect any flower in the neighborhood, steer the drone towards the flower and make the drone's pollinating rod touch the flower. We trained all deep learning models based on an artificial dataset with a mix of images of real flowers, artificial (synthetic) flowers and virtually rendered flowers. Our experiments demonstrate that the approach is technically feasible. The drone is able to detect, approach and touch the flowers totally autonomously. Our 10 cm sized prototype is trained on sunflowers, but the methodology presented in this paper can be retrained for any flower type.

Keywords: pollination drone; visual servoing; two-stage approach; neural network

1. Introduction

World food consumption is projected to increase in the coming decades [1]. Together with a declining insect population [2], and given the roles bees play in the pollination of crops [3], this is a big concern for world food security. To have a successful crop, a farmer might rely on the surrounding ecosystem, artificially introduce a bee colony to pollinate their crops or rely on some other biological pollination method. However, due to collapse of surrounding ecosystems, regulatory difficulties such as protection from invasive species, and the circumstances of an artificial factory environment, interest in human made pollination methods has grown. Currently, pollination robots are already being used [4,5].

These methods use mobile robot arms together with some kind of pollination brush to go from flower to flower, distributing pollen from plant to plant. Using such a robot arm however does have some limitations: (i) the way a farm is laid out might for instance not easily allow a robot to pass between different crops, (ii) the type of produce might not lend itself to be touched by a robot arm, or (iii) the field might be too big or steep to build infrastructure for such a mobile ground robot. Presently, some crops are even planted above each other in vertical farms, which makes it impossible for a wheeled robot to pollinate [6]. All of this means that there is a lack and a certain need for a more generic and versatile pollination method.

In this paper, we present work towards a drone equipped with a camera and on-board processing to autonomously pollinate flowers. The approach we present enables a drone to autonomously detect flowers and approach them using a two-stage deep learning approach.

As a preliminary use case, we focus here on sunflowers. This is a deliberate choice, as our final goal would be to develop a nanodrone (with a size of a few cm), that can pollinate small crop flowers, such as strawberry flowers, cucumber flowers, etc. As the hardware at such small scale is not obtainable yet, in this work we demonstrate the technology at a slightly larger scale, where the large size of sunflowers matches with the scale of the 10 cm mini-drone used.

The three main novelties of this paper are:

1. We demonstrated the deployment of deep learning based computer vision in real-time on-board a resource restricted embedded processing platform.
2. We developed a methodology to train the necessary neural networks with a partially real, partially synthesized dataset.
3. We successfully demonstrated a two-stage flower approaching visual servoing procedure.

In the remainder of this paper, we investigate how to create such a visual navigation system for a pollination drone. In Section 2, we lay out the related work on robot pollination methods, autonomous drone technology and the computer vision techniques we use to steer the drone. In Section 3, we go in depth into the inner workings of our drone navigation approach. In Section 4, we explain the experiments we did to evaluate the effectiveness of our approach and discuss our promising results. We conclude in Section 5. This paper is an extended version of the conference paper [7] we previously presented at ARCI 2022.

2. Related Work

In this section we will first present the basics of drone navigation and visual servoing. Then we investigate existing artificial pollination solutions and then go into hardware platforms that can be used for on-board computer vision based navigation.

2.1. Drone Navigation

In this section, we will briefly discus the Degrees Of Freedom (DOF) of our drone and its main components. In general, a drone has four or more propellers that deliver thrust to keep the drone into the air. When all propellers deliver an equal amount of thrust and generate an upward force equal to gravity, the drone is hovering. This means that the drone maintains its position in the air. To fly in a certain direction, some of the propellers are slowed down causing a rotation and a translation in that direction. When all propellers spin in the same direction, the drone would start rotating around its own Z-axis. To counter this torque effect, two of the four propellers turn clockwise and the other two turn counter-clockwise as in seen in Figure 1 (left).

The drone that we are using for this paper is a DJI Tello Robomaster TT. This type of drone has six DOFs as seen in Figure 1 (right). When the front and back pairs of rotors (1&2 vs. 3&4) are steered to turn with a different speed, the downward thrust vector turns around the Y-axis. This results in a horizontal force component, causing the drone to translate forward or backwards. This movement is called the *pitch*. A rotation around the X-axis (slowing down the left or right pair of motors) yields a translation to the left or right, called the *roll*. The roll and pitch are directly coupled with a translation left, right, forward and backward. When a diagonal motor pair gets another speed w.r.t. the other pair, a net rotation momentum causes the drone to rotate around its vertical axis. This is called the *yaw*. The Z-axis can be controlled separately by giving all four motors the same speed increase or decrease, to adjust the altitude of the drone.

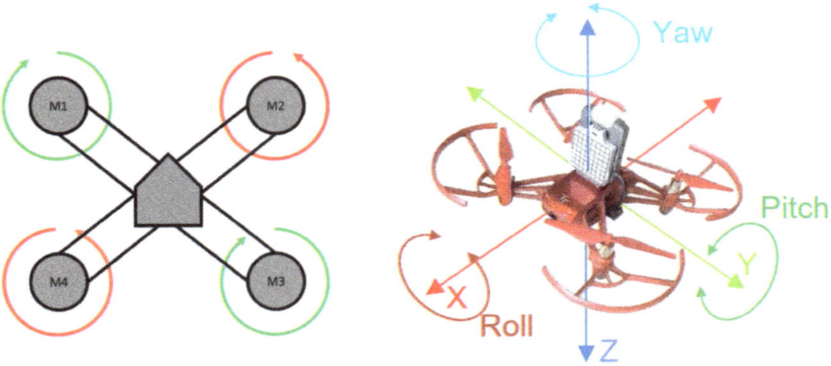

Figure 1. (**Left**): A quadrocopter with the directions of the turning propellers displayed. (**Right**): The six DOFs of our drone.

As seen in Figure 2 the heart of a drone is its flight controller. This controller ensures that the drone remains in an idle horizontal position when no control commands are sent to the flight controller, called *hovering*. When control commands are received, the flight controller calculates a proper speed for each of the (in our case four) motors, e.g., when a "go forward" command is received, the flight controller lowers the speed of the two motors in the front which yields a forward movement. Without this flight controller, the pilot should control the four motors independently to make this movement, which is almost impossible. To stabilize the drone, the flight controller tries to maintain its angle in the X, Y and Z plane. To do this, the current orientation of the drone is calculated using two main sensors as explained below.

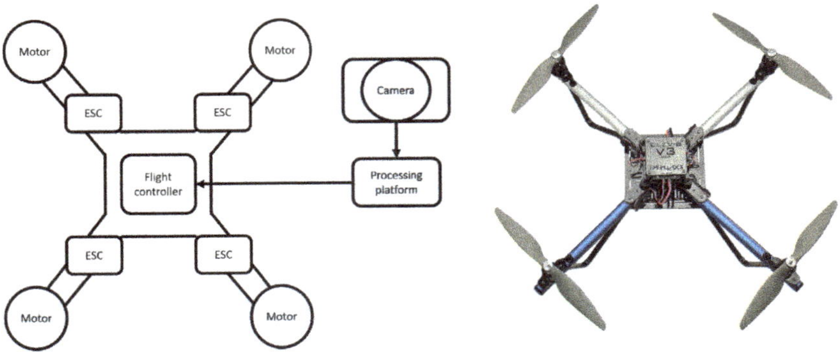

Figure 2. The main components of a drone: Flight controller, ESC (Electronic Speed Controller) and motors. We added an extra camera and processing platform to our drone.

The first is the accelerometer, which outputs the acceleration in three directions (for X, Y and Z). This acceleration is expressed in gravity units (1 g = 9.8 m/s^2). Thus, when the acceleration sensor is not moved, the output will be 1g for the z-axis and 0g for the X- and Y- axis of the sensor. These values are used to calculate the angle of the sensor in three directions but cannot be used directly to measure the orientation of the drone. The downside of an acceleration sensor is that the output is not reliable on its own. The accelerometer measures inertial force, such a force can be caused by gravitation (and ideally only by gravitation), but it might also be caused by acceleration (movement) of the device. Thus, when the drone is moving, the output of the accelerometer cannot directly be used to capture the gravity vector. Furthermore, even if the accelerometer is in a relatively stable state, it is still very sensitive to vibration and mechanical noise in general. To deal with these problems, a gyroscope is used to smooth out these errors. A gyroscope measures the rate of changes of the angles in deg/s which is less sensitive to linear mechanical movement. Although, the disadvantage of a gyroscope is that it suffers from drift (not coming back to zero when rotation stops). Nevertheless, by fusing the data of both sensors, an accurate angle of the drone can be obtained to help the flight controller stabilize the drone.

Our flight controller also contains a barometer to maintain altitude and a small down-looking camera to maintain a fixed position. The accuracy of these sensors is of less importance in our experiments, since we use Image-based Visual Servoing which does not use global coordinates as further explained. When the flight controller has calculated a velocity for each motor, the velocity is passed to a Electronic Speed Controller (ESC) which drives the motors. As shown in Figure 2 we also equipped our drone with a camera to capture images which are used by the algorithm running on the embedded processing platform. Our task is to develop software that interprets the images and generates control commands for the flight controller to steer the drone. This software should run in real-time on an embedded platform, mounted on-board.

2.2. Visual Servoing

Visual servoing is the technique we use to steer our drone, using information from the camera. This technique can also be used to control e.g., a robot arm, where the camera can be mounted on the end effector of the robot arm (eye-in-hand), or somewhere else fixed on the frame (eye-to-hand). In the case of a drone, the camera is fixed to the drone itself, and thus an eye-in-hand system. There are two main visual servoing techniques: Position-based Visual Servoing (PBVS) and Image-based Visual Servoing (IBVS).

In PBVS, the 3D position of the object (in our case a flower, in the case of a robot arm an object to grasp) is calculated w.r.t. the camera. Then, a control command is calculated to go from the current 3D position of the end effector to the goal position. When the object is moving, its new 3D position and the 3D position of the end effector should be estimated in every frame, to recalculate a new path to move. The disadvantage of this approach is that the accuracy of the system is sensitive to calibration errors.

In IBVS, only the 2D position of the object is estimated in the frame and used to calculate an error between the current 2D position and the desired 2D position for every frame. This error can then be used to control the actuator, or drone. We use IBVS in this paper, since our camera is mounted on the drone itself and we want to position our drone w.r.t. the flower. The 3D position in the world is of less importance. Indeed, our main goal is to approach the flower, even if it moves slightly. A more detailed description of PVBS and IBVS can be found in [8].

2.3. PID Control Loops

In this subsection, we illustrate the control loop we used to steer the drone smoothly with an example. The drone can be steered in different ways to approach a flower. In this example, we will only use the pitch to position the drone at a fixed distance, e.g., 4 m, w.r.t. a flower. When the distance between the drone and flower increases, the pitch should be controlled to fly forward and vice versa. When controlling the pitch, the drone will rotate around its Y-axis (see Figure 1) with a certain angle resulting in a net horizontal force and hence a translation of the drone. The flight controller enables us to simply send the desired velocity of the translation, which is then automatically recalculated to a certain angle for the pitch.

A possibility is to control the pitch binary such that the pitch velocity is maximal, e.g., 10 m/s when the distance is larger than 4 m, and −10 m/s when the distance is smaller than 4 m. Indeed, this will result in a very jerky movement of the drone. A better way to control the pitch is proportional to the error between the desired distance and the current distance as in Figure 3. Here, the measured distance is 4.2 m, resulting in an error of 0.2 m.

This error is multiplied by a proportional (K_p) gain (e.g., 5 s^{-1}) resulting in a pitch velocity of 1 m/s. Using only the K_p gain, the control loop is formulated as in this equation:

$$v = K_p \cdot E(t) \quad (1)$$

where v is the velocity and $E(t)$ the error.

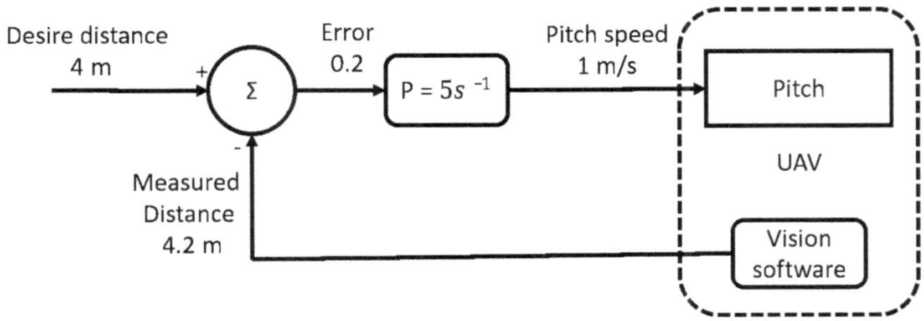

Figure 3. A proportional control loop. The error between a desired and measured value is calculated and multiplied with a scale factor to generate a control value.

By increasing or decreasing the K_p, the desired distance can be reached faster or slower. So, why not making K_p as high as possible, e.g., 100 s^{-1}? When K_p is high, the velocity of the pitch will also be high, even at a small error. An error of e.g., 0.1 m will result in a velocity of 10 m/s and the drone cannot be stopped in time (when reaching 4 m) due to its momentum of inertia, and will overshoot and oscillate around its desired distance. A high proportional gain can even lead to an unstable system, where the output of the P controller increases every frame until the maximum speed of the drone is reached. We indeed better limit the maximum velocity of the drone to a safe value. Making K_p smaller can reduce the oscillation, but it will take longer to decrease the error and it is even possible that the drone cannot reach its desired distance. When P is small, e.g., 0.1 s^{-1} at 4.2 m, the error will be 0.2 m and the velocity 0.02 m/s. Such a small velocity will not move the drone enough and the desired distance cannot be reached.

To overcome this problem, the integral term (I) is introduced as in Figure 4. The I term integrates the error over time. When the error is too small to move the drone, this error will remain and the I term sums this error over time. This means that the longer the error exists, the higher the output of the I term will be, resulting in an increase in velocity, which eventually will move the drone until the error becomes 0. The control loop can now be formulated as:

$$v = K_p \cdot E(t) + K_i \int_0^t E(t) \cdot dt \quad (2)$$

The velocity v is the sum of the P and I terms.

As mentioned before, we want our K_p gain as high as possible to react as fast as possible. A high K_p gain induces overshoot. To damp this overshoot a derivative (D) term is introduced as in Figure 5. This D term will look at the rate of change of the error over time. When the drone comes closer to its goal distance of 4 m, the error will become smaller very fast. The D term senses this fast change and starts to generate an inverse control value, to slow down the drone even faster (braking), resulting in a minimal overshoot. The final PID control loop is formulated as:

$$v = K_p \cdot E(t) + K_i \int_0^t E(t)dt + K_d \cdot \frac{dE(t)}{dt} \quad (3)$$

The final velocity v is the sum of the P, I and D terms.

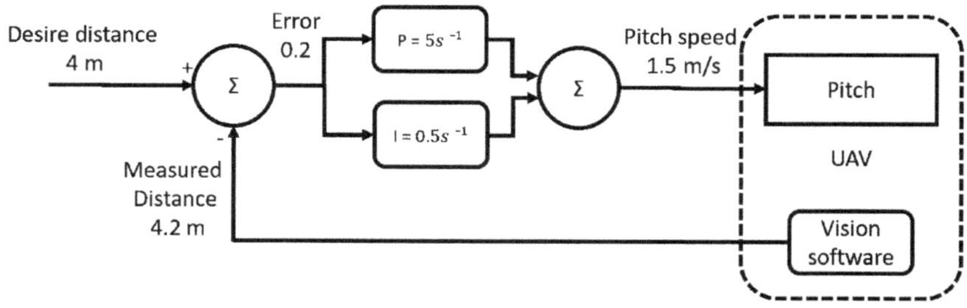

Figure 4. A proportional-integral control loop. The integral factor ensures that the error can decrease to 0.

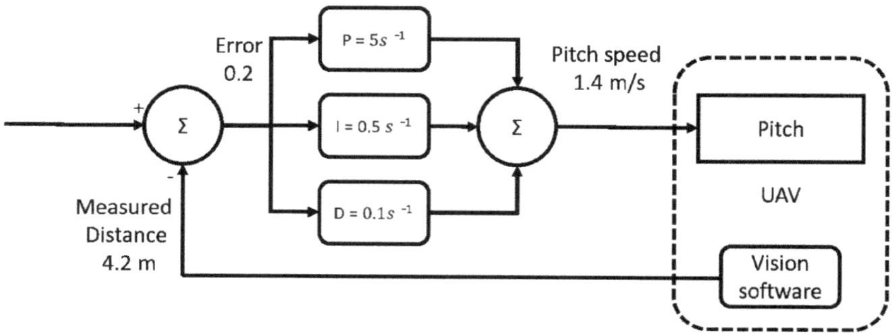

Figure 5. A proportional-Integral-Derivative control loop. The D factor damps the overshoot induced by P and I.

For each of the axes, such a PID loop is implemented to control the drone smoothly. Of course, K_p, K_i and K_d are tuned for each axis as explained in Section 3.4.1.

A PID control loop is a closed-loop controller, since the action from the controller is dependent on the process output. This can also be combined with an open-loop controller (feed-forward controller). Here, knowledge of the inertia and acceleration of the drone is fed forward in the control loop to make the output even more stable. We did not use this combination, since there is no inertia model of our drone publicly available. However, in the future, this model can be approximated doing real-life experiments.

PID control loops are not the only control loops that can be used to steer a drone. In [9], Fuzzy Logic was used to control a drone to fly waypoints. The advantage of using Fuzzy Logic is that the reaction of the controller on a certain input can be formulated more precisely for different states of the system. Although, since different states of the system can be tuned separately, a large number of parameters should be fine-tuned compared to a PID loop. Hence, for most systems, a PID control loop is preferred above Fuzzy Logic.

Another approach is to use iTaSC (instantaneous task specification and control) [10] as a control loop. iTaSC is a systematic constraint-based approach to specify complex tasks of general sensor-based robot systems, such as a drone. The advantage of using iTaSC is that it can automatically derive controller and estimator equations from a geometric task model, that is obtained using a systematic task modeling procedure. We did not use iTaSC in this paper, due to the more time-consuming integration w.r.t. PID control loops. Nevertheless, it would be interesting to use this approach in the future. Indeed, iTaSC can control several axes of the drone automatically to accomplish a specific movement using a geometric model of the drone, where in our approach all axes should be controlled independently by different control loops.

2.4. Related Work on Artificial Pollination

As mentioned earlier, world food consumption is projected to increase significantly in the coming decades [1]. To address this challenge, more efficient ways of producing food should be developed to feed this growing demand. A possible solution using technological means is to aid nature with the pollination of crops and plants. In recent research, robots were already developed to automatically pollinate crops and flowers by using a robotic arm on a base equipped with wheels [5,11]. For the detection of the flower, they use Inception-V3 [12] together with color segmentation. This is a common approach for detecting flowers as seen in [13] where they combine a CNN and SVM to predict an even more accurate segmentation of the flowers. The downside of a robotic arm on a wheeled platform is its size and maneuverability.

In [14] a drone was used to pollinate the flowers. This drone has multiple advantages over a wheeled robot, but was way too big (50 cm × 50 cm) to do the extremely precise job of pollinating flowers. In this work we developed a small drone (10 cm × 10 cm) with on-board processing power which can estimate the position and angle of the flower. Furthermore, this drone can autonomously fly towards the flower in order to precisely pollinate the flower.

2.5. Hardware Platforms for On-Board Computer Vision Based Navigation

We investigated many hardware platforms that would fit our criteria for on-board computation of deep learning based computer vision tasks. To meet our requirements, the platform should be: (i) light-weight, (ii) low-power, (iii) easy to deploy and integrate with an existing drone, and (iv) able to run state-of-the-art computer vision algorithms with a sufficiently high frame rate.

For the most part, this rules out higher power mini-computer devices such as the NVIDIA Jetson series. The STM32F746 or other ARM Cortex M7 based platforms provide a deep learning library (CMSIS-NN) to run deep learning applications on the CPU architecture. The Ambiq Apollo3 Blue development board is able to run Tensorflow Lite models mostly targeted at voice and gesture recognition, but not computer vision. The Greenwave GAP8 is a RISC-V processor optimized to run deep learning models in a multi-threaded way. The Kendryte K210 processor is also based on a RISC-V core, but in addition to that also contains a deep learning accelerator that is able to run YOLO [?] models at up to 20 fps. Due to its performance and relative light weight, we decided that the Kendryte processor best fits our use case.

From a machine learning point of view, we need models that are able to do two types of tasks: detection and classification (direct steering). These areas are a very active field of research, and many pipelines are available, each with their own trade-offs. Discussing all current techniques would lead us too far from the scope of this paper. Instead, we will discuss the techniques that are available on the chosen platform here. The Kendryte processor has support for the Mobilenet [16] family of backbones. This is a backbone especially optimized for low-power mobile devices, and is able to save a tremendous amount of computing power by using depthwise convolutions. Since the publication of the Mobilenetv1 paper, many additions to the network have been proposed as well [17]. However, due to the platform we choose, we are limited to Mobilenetv1. The classification head uses a simple cross entropy loss. In the detection stage we use a modified detection network, together with a Mobilenet to do fast inference. The detector makes bounding box predictions with only a single pass over the network, as opposed to region-based detectors like Faster-RCNNN [18] and Masked-RCNN [19]. Due to their nature, generally speaking, single stage detectors are faster than region-based ones. The authors in [20] did a complete survey on the current state of object detection algorithms.

3. Autonomous Drone Navigation

In this section we go further into detail about the inner workings of the visual navigation system for our pollination drone. We create an artificial sunflower dataset containing images of real, as well as artificial and virtual sunflowers. We use the low power Kendryte K210 hardware platform which is able to run quantized Mobilenet based networks mounted on a commercially available DJI Tello Talent drone. A two stage flower approaching technique is developed to successfully touch the flower. In the first stage, we detect the flower from a distance between 8 and 0.8 m using a convolutional detection network. In the second stage (0.8–0 m) we use an end-to-end network to predict the center of the flower in the camera feed. To guide the drone towards the center of the flower, a PID control loop is used to convert the location of the flower to steering commands for the drone. In this section, we explain all of these components one by one.

3.1. Flower Dataset

To train the computer vision neural networks, we created an artificial dataset based on images of real sunflowers, artificial (synthetic) sunflowers and virtually rendered sunflowers. An important aspect we take into account is that for the drone to successfully pollinate a sunflower, it should approach the sunflower in a line perpendicular to the sunflower. In order to be able to do this, we need to estimate the angle of gaze orientation of the sunflower to enable us to correct for this when we are planning the drone's trajectory. This means that in each image the angle of the flower w.r.t. the camera should also be annotated.

For the virtual flowers, we simply render them at different angles, both in horizontal and vertical direction. For the real and synthetic sunflowers, we made a recorder device that is able to rotate the flower automatically over these two degrees of freedom while taking pictures of the flower. Both are illustrated in Figure 6. The generated images are then pasted on images from the Places dataset, more specific on images from Japanese gardens [21] to use as background. A total of 5660 annotated images with multiple flowers, and 4500 zoomed-in images on one flower were generated this way. Examples of these images can be seen in Figure 7.

Figure 6. Acquisition of the training data. Top: 3D models of flowers are rendered at different angles and positions. Bottom: our chroma key set-up that is used to acquire images from both artificial (**left**) and real (**center**) sunflowers, with the drone's camera, using a 2 DoF rotator device (**right**).

3.2. Hardware Platform

Since the precise steering the drone has to perform, processing of the video feed should happen on-board such that the delay between receiving images and correcting the drone's position is minimal. This implies the need for a small and light-weight processing board capable of performing real-time image processing. We choose to use a Maix Bit development-board which contains a Kendryte K210 (RISC-V) processor. This board measures 53 mm × 25 mm and weighs 25 g. Furthermore, this board is equipped with a camera, resulting in a minimal image receiving delay.

Figure 7. (**Left**): 3D generated sunflowers for the flower detector. (**Right**): Generated images for direct visual servoing approach.

The camera feed is directly read by the Maix Bit on which we either run the end-to-end network or the detection network. The Maix Bit communicates with the Tello drone via its on-board ESP32 microcontroller. In addition to the camera, we also have a ToF depth sensor (VL53L1X) to measure the distance from the flower in the final approach. A detailed overview of our setup is shown in Figure 8. Figure 9 shows the drone carrying the Maix Bit.

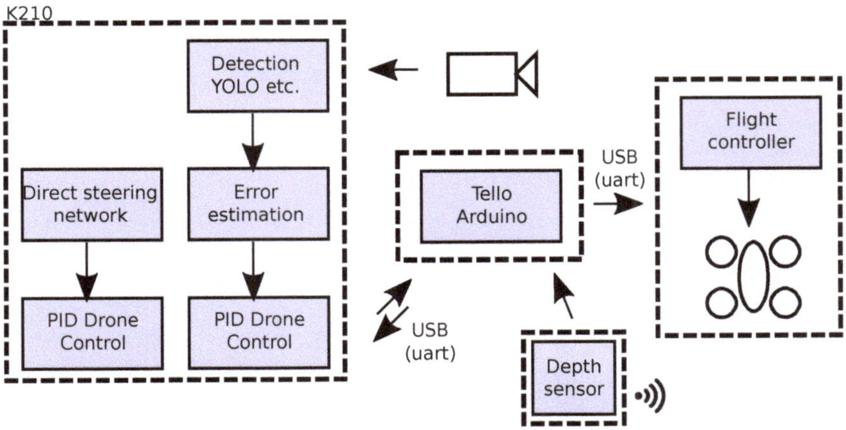

Figure 8. Overview of our hardware setup.

3.3. Hybrid End-to-End and Detection Approach

Figure 10 gives an overview of the piece-wise visual servoing approach we developed to steer the drone towards a flower. As explained before, the vision part of our approach plan is divided into two stages. Figure 10 gives an overview of our visual servoing approach. In the first stage we run a CNN for flower pose estimation: position, size (indication of the distance) and orientation. This model steers the drone to fly towards a position close to the flower (approximately 80 cm), directly facing it. This stage is done in three steps: (i) a forward movement until the drone is at the right distance from the flower, (ii) an altitude change such that the drone is at the flower's height and (iii) an encirclement motion around the flower, until the drone faces the flower directly. In the second stage, we used an image-based visual servoing network in an end-to-end approach such that it directly outputs steering commands towards the sunflower's position. This model is used for the final approach, the pollination touchdown.

Figure 9. Our drone prototype carrying the Maix Bit.

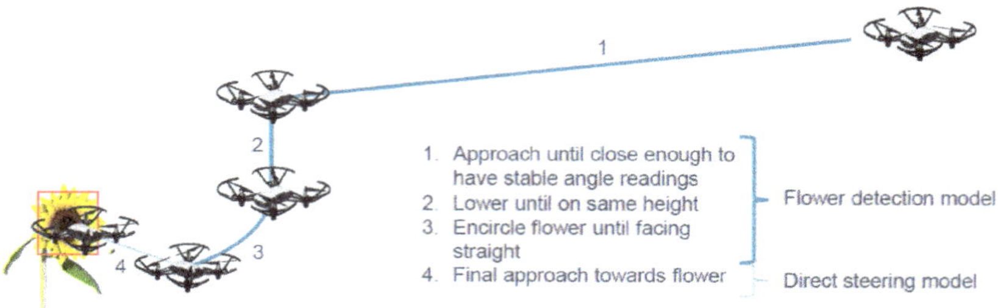

Figure 10. Piece-wise visual servoing approach overview.

3.3.1. Detection Stage

In the first stage (distance of 8 m to 0.8 m) we use a MobileNet [16] detector trained on our own flower dataset. The position of the flower is used to steer the drone such that the flower is always in the center of the drone's view. In this stage, the size of the detected flower is used as a distance measurement. Figure 11 illustrates the measured properties of the flower that our detector outputs, and the steering dimensions associated with them.

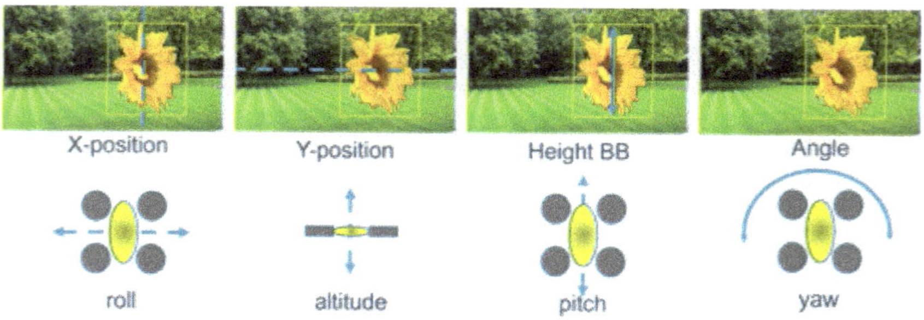

Figure 11. The four properties that our detection model perceives of the flower and their related steering dimension.

As indicated, we altered the CNN architecture to additionally output the horizontal angle of the flower, illustrated in Figure 12. This predicted angle is used to steer the drone towards a 0° angle w.r.t. the flower. When the drone approaches the flower and the size of the flower exceeds 60% of the frame-height, the flower becomes too big to be reliably detectable, and we switch to the second visual servoing stage.

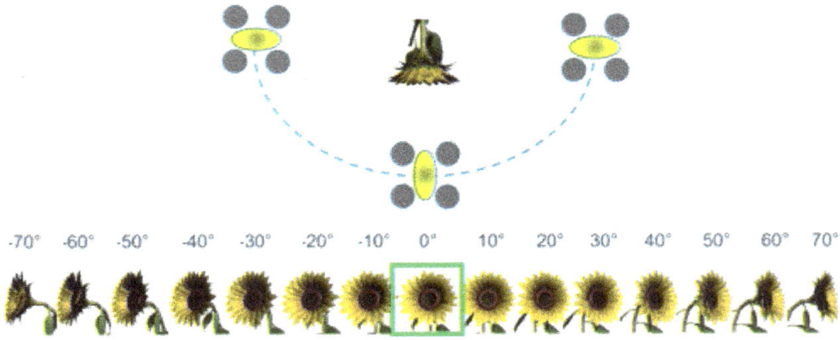

Figure 12. Illustration on the horizontal angle detection of the flower.

3.3.2. Direct Visual Servoing

When the drone approached the flower close enough for the final descent, our detection model of the previous phase does not work anymore. That is because the flower's size exceeds the field-of-view of the camera. In this stage we use an end-to-end network based again on a Mobilenetv1 architecture trained for classification that directly outputs steering commands (up, down, left, right or center). This network is trained on zoomed-in images of a flower (see Figure 13). During the final approach to the flower, we use a VL53L1X ToF distance sensor to measure the distance between the drone and the flower with high accuracy. The pollination rod in front of the flower measures 8 cm. When the distance becomes smaller than 8 cm we assume the rod touched the flower and the pollination took place.

Figure 13. (**Left**): example images from the direct visual servoing image dataset. (**Right**): The proposed direct visual servoing approach.

After pollinating the first flower, we plan to let the drone do a random maneuver (consisting of a backward and upward movement, random horizontal rotation and an adjustable horizontal movement). Then, the drone will be in a new random position from where it can start again approaching the closest sunflower. This random motion will eventually make it possible to visit multiple flowers in a field.

3.3.3. Optimization of the Neural Networks

In order to run neural networks on a resource-restricted embedded device, optimization is necessary. From a memory point-of-view, the biggest challenge is storing all model weights locally. This is essential to guarantee real-time execution. However, also the activation map memory storage and the amount of compute operations needs to be restricted, even for an already very efficient and lightweight model as the MobileNet architecture we use for both models. To cope with this, we opted to post-quantize the parameters of our TFLite model from 32 bit floats to 8 bit ints. Additionally, the NNCase compiler included within the MaixPy Kendryte toolchain optimized our models further using operator fusion and memory optimization. This enabled us to squeeze both MobileNetV1 models together into the 8 MB RAM of the tiny Kendryte K210 hardware platform we use.

3.4. PID Control Loop

To transfer the position of the flower into steering commands for the drone, four different PID loops are used, one for each of the drone's axes. The goal of the PID loops is to center the flower in the drone's view and steer the drone such that the angle between the drone and the flower reaches zero degrees (ideal for pollination). To position the drone such that the flower is in the center of the screen, the Y-coordinate of the flower is used to move the drone up or down (Altitude) while the X-coordinate is used to steer the drone to the left or right (Roll). The angle of the flower is used to control the drone's rotation around its Z-axis (Yaw). When the Yaw is changed, consequently, the flower will also move to the left or right in the image, which will be again compensated by controlling the Roll. To approach the flower, the size of the detection is used to move the drone forwards or backwards (Pitch). When the distance becomes smaller than 0.8 meter, we switch to an end-to-end stage where the time of flight distance sensor is used to control the Pitch instead.

3.4.1. Tuning of the PID Loops

Each axis of our drone is steered by a control loop that gets its information from the vision software. In this section we explain how we tuned K_p, K_i and K_d for each control loop. A first tuning of the PID parameters happened in simulation. DJI provides a simulator to learn how to fly with their drones and the (virtual) drone in the simulator reacts very similar to the real one, since DJI made a dynamic model of the Tello. Since we cannot insert flowers and a camera in the simulation environment, we replaced the flower with a specific GPS position and developed software to fly to that GPS position and stop when it arrives. First K_i and K_d were set to 0 and K_p was chosen so that the drone was just not oscillating around the goal GPS position. Then K_i was increased to reach the GPS position (make error 0), with the result of a small overshoot. Next K_d was increased to damp this overshoot. Of course, since our software induces small delays while measuring the depth, the position of the flower and the angle of the flower, the PID values are also fine-tuned in real-life. For this we developed an android-app to tune the PID values in real-time. Nevertheless, this fine-tuning took a lot of time since each time a specific loop changed, its PID settings had to be adjusted. To further overcome oscillations around the setpoint we inserted a dead-zone in our system. This is a zone around the setpoint where the drone should not correct its position. A threshold value for the corrections was also chosen empirically. These PID control loops run at a fixed speed of 50 Hz, all in separated threads, which is fast enough to provide the flight controller with data.

4. Experiments and Results

In this section, we will summarize the experiments we have conducted with the developed visual navigation system.

4.1. Flower Detection

The foremost task of the on-board vision processing is detecting the flower in the image, such that the drone can steer itself towards it. For this, we implemented the MobileNet flower detector. Figure 14 shows some qualitative detection results, both when multiple flowers are in view, and at various distances. When multiple flowers are in view of the drone, it selects the largest and most confident flower detection to first navigate towards.

Table 1 shows a comparison of different models we trained for the combined task of flower detection and angle estimation. For there models, we started from pretrained weights on ImageNet, and trained for 50 k steps. The alpha factor controls the width of the network. This is known as the width multiplier in the MobileNet paper. If alpha < 1.0, it proportionally decreases the number of filters in each layer. The detection result is evaluated using mAP, the mean average precision. For each of the detection thresholds, we evaluate how many detections overlap enough (using an intersection over union threshold of 0.5) with ground truth bounding boxes. These precision results are averaged over the detection thresholds, yielding the AP (average precision) score for each class. After optimization of the models (see Section 3.3.3), we chose the MobileNetV1 model with alpha factor 0.5, because that is the best performing model that fits in the hardware.

Figure 14. Illustration of detection results both when there are multiple flowers in view (**upper**), and when there is only one flower in view (**lower**).

Table 1. Experimental results of different models for flower detection and angle estimation.

Model	Alpha Factor	Resolution	Detection AP
SSD + MobileNetV1	1.0	320 × 240	0.68
SSD + MobileNetV1	0.5	320 × 240	0.61
SSD + MobileNetV1	0.25	320 × 240	0.51
SSD + MobileNetV2	1.0	320 × 240	0.77
SSD + MobileNetV2	0.5	320 × 240	0.66
SSD + MobileNetV2	0.35	320 × 240	0.62

On the real-world evaluation dataset, we evaluated the average precision of the angle detection results of our chosen model. As explained above, our model uses 10° angular bin classes for angle detection. The final mAP is computed as the mean AP over all these classes. We reached a mAP of 0.36 on this test dataset (note that the different angle classes in our dataset make our dataset more challenging; when doing real-world tests this accuracy proved sufficient).

In Figure 15 we evaluate the detection accuracy against the distance the flower is moved from the camera (equivalent to the size of the flower in the image), we can see that flowers further away from the camera are harder to detect. Moreover, if flowers are too close by, the detection also begins to fail. That is the main reason for the second model (direct steering) model we propose, where no explicit object bounding boxes are computed.

4.2. Direct Steering Model

In the second phase, when the drone is close enough to the flower, the navigation is taken over by our second computer vision model. As described before, this model directly outputs motion commands, based on the centeredness of the (visible part of the) flower in the camera image. First, we trained a MobileNetV2 architecture on our direct steering training dataset (see Figure 13). Unquantized, using Float32 number representations, this model reaches 86.2% accuracy, after quantization to Int8 70.0%. A MobileNetV1-model yields 70.5% in Float32 and 69.7% in Int8. On the *difficult* test part of our direct steering dataset, this MobileNetV1 model achieved 53% accuracy.

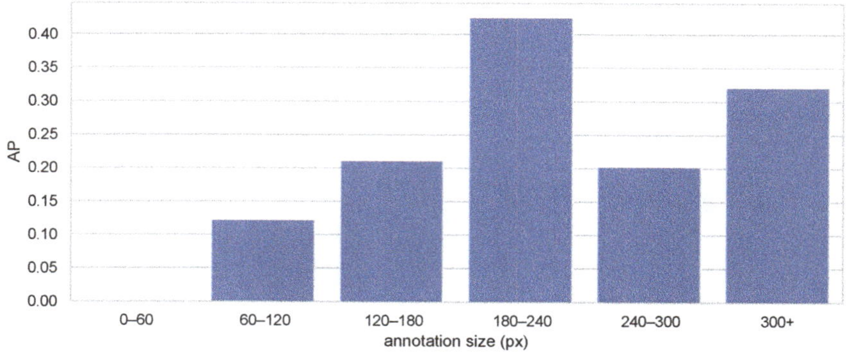

Figure 15. Influence of the distance from the sunflower (smaller flowers are assumed to be further away from the camera) on the detection accuracy.

In order to test this model separately, we wrote a control loop that steers the drone such that it keeps facing the flower directly while maintaining a fixed distance of 20 cm. This intensively uses our on-board running direct steering model, and additionally the time-of-flight sensor for the distance measurement. A demonstration of this experiment can be seen at https://youtu.be/oxx7KPLV804 Accessed on (9 May 2022). It is clear that the accuracy of the direct steering model is sufficient to steer the drone accordingly.

4.3. Visual Navigation

In order to validate the full system, we conducted an experiment evaluating the success-rate of steering a drone towards the pollinating position for a sunflower. We repeated this experiment 24 times where the drone took off from a distance of 5 m to 8 m, flew to the flower, touched it and flew back. In total, we reached a success-rate of 87%. In the other 13% of non-success, a false detection was the cause of failure. For the successful navigation trials, the average time to reach the pollination position is 31.5 s, with a standard deviation of 15.3 s.

The maximum frame rate of which images could be processed on the Maix Bit is approximately 30 fps for the direct steering model and 20 fps for the detection network. We also tested the minimum frame rate needed to smoothly steer the drone towards the flower since a slower frame rate results in less power consumption. The minimum framerate needed to smoothly approach a flower is measured at 12 fps.

Table 2 provides URLs to videos of six of these trial flights. Under different light conditions, our drone was able to successfully navigate towards the flower. Sequence 3 also demonstrates that our system can cope with outdoor conditions. However, because of the small weight of the drone, we notice that too much wind is certainly a challenge for precise navigation.

Table 2. Demonstrations of the full navigation approach.

Sequence nb	Conditions	Remarks	Video URL (Accessed 9 May 2022)
1	indoor	cluttered environment	https://youtu.be/quX5HhVyR3g
2	indoor	demo at Dubai World Expo	https://youtu.be/u13j3sPgDlE
3	outdoor	low wind conditions	https://youtu.be/ixOCjHggUw4
4	indoor	difficult light conditions	https://youtu.be/DZh7zHVQJqM
5	indoor	initial heading away from flower	https://youtu.be/Lq7TR70cJJk
6	indoor	long search for flower	https://youtu.be/AhhI29ofmr0

In earlier experiments (e.g., sequences 1 and 2 in Table 2), we noted that the final approach towards the flower is functional, but that it centers the camera with the center of the flower, and not the pollination rod we mounted on it. In later experiments (e.g., sequences 3–6), we corrected this vertical bias, such that the pollination rod is now exactly steered towards the flower's center. Figure 16 shows four of these successful trials. As seen in the bottom row, the drone's pollinating rod exactly touches the center of the flower now.

Figure 16. (Columnwise) Successful navigation trials towards the pollination position.

4.4. Discussion

In this paper, we focused on steering an autonomous drone towards a flower for pollination. We know that in reality, a successful pollination occurs only when a bee lands on the head of the sunflower and drags pollen from its legs onto the stigma in the head of the plant. Just touching part of the flower will not induce pollination. Therefore, in order to extend the current prototype towards a real pollination device, it needs a pollination tool that has the same effect of this biological action, such as a brush, vibrating rod or even a pressured air gun. With our simple pollination rod, visible in the demonstration videos, successful sunflower pollinations are impossible. However, the developed navigation system is necessary in any case, including for other pollination implements.

5. Conclusions

In this paper we developed the navigation system for a pollination drone that detects flowers, estimates the angle and flies towards the flower in a two stage approach. Different experiments were performed to evaluate all the parts of this approach, yielding promising results. We managed to run all processing on-board, resulting in a fully autonomous drone with a minimal delay between taking images and the control of the different degrees of freedom. Real world test demonstrate a navigation success rate of 87%.

Furthermore, in this work we demonstrate that, using currently available hardware, it is possible to autonomously fly a drone towards a flower and maneuver it in a way similar to what a bee would do to pollinate a flower. Although the present prototype is trained for sunflowers, the methodology can easily be retrained for any other flower species based on our hybrid training approach with simulated and real images.

However, it remains to be seen how well methods like these would scale up to replace actual bee populations, and what other environmental impacts a dwindling insect population can have on the survival of humanity, and the surrounding ecosystems it relies on. A discussion should be had on how much we want to rely on technological solutions for our own food security, and what role humanity should play in interfering in existing ecosystems.

Author Contributions: Conceptualization, D.H., W.V.R., Y.C., T.G.; methodology, D.H., W.V.R., Y.C., T.G.; software, D.H. and W.V.R.; validation, D.H. and W.V.R.; investigation, D.H., W.V.R., Y.C. and T.G.; resources, Y.C., T.G.; data curation, D.H. and W.V.R.; writing—original draft preparation, D.H., W.V.R. and T.G.; writing—review and editing, D.H., W.V.R. and T.G.; visualization, D.H., W.V.R. and T.G.; supervision, Y.C., T.G.; project administration, Y.C., T.G.; funding acquisition, Y.C., T.G. All authors have read and agreed to the published version of the manuscript.

Funding: This work has been supported by the company MAGICS, the FWO SBO project OmniDrone under agreement S003817N and the Flemish Government under the AI Research Program.

Data Availability Statement: Due to the commercial nature of this research we are not able to make the sunflower dataset publicly available. However the methods described in this paper could be used to create a dataset with similar properties.

Conflicts of Interest: The authors declare no conflict of interest.

References

1. Van Dijk, M.; Morley, T.; Rau, M.L.; Saghai, Y. A meta-analysis of projected global food demand and population at risk of hunger for the period 2010–2050. *Nat. Food* **2021**, *2*, 494–501. [CrossRef]
2. Wagner, D.L. Insect declines in the Anthropocene. *Annu. Rev. Entomol.* **2020**, *65*, 457–480. [CrossRef] [PubMed]
3. Khalifa, S.A.; Elshafiey, E.H.; Shetaia, A.A.; El-Wahed, A.A.A.; Algethami, A.F.; Musharraf, S.G.; AlAjmi, M.F.; Zhao, C.; Masry, S.H.; Abdel-Daim, M.M.; et al. Overview of bee pollination and its economic value for crop production. *Insects* **2021**, *12*, 688. [CrossRef] [PubMed]
4. Yuan, T.; Zhang, S.; Sheng, X.; Wang, D.; Gong, Y.; Li, W. An autonomous pollination robot for hormone treatment of tomato flower in greenhouse. In Proceedings of the 2016 3rd International Conference on Systems and Informatics (ICSAI), Shanghai, China, 19–21 Noember 2016; pp. 108–113.
5. Ohi, N.; Lassak, K.; Watson, R.; Strader, J.; Du, Y.; Yang, C.; Hedrick, G.; Nguyen, J.; Harper, S.; Reynolds, D.; et al. Design of an Autonomous Precision Pollination Robot. In Proceedings of the 2018 IEEE/RSJ International Conference on Intelligent Robots and Systems (IROS), Madrid, Spain, 1–5 October 2018; pp. 7711–7718. [CrossRef]
6. Kalantari, F.; Mohd Tahir, O.; Mahmoudi Lahijani, A.; Kalantari, S. A review of vertical farming technology: A guide for implementation of building integrated agriculture in cities. In *Advanced Engineering Forum*; Trans Tech Publications Ltd., Switzerland: Bach, Switzerland, 2017; Volume 24, pp. 76–91.
7. Hulens, D.; Van Ranst, W.; Cao, Y.; Goedemé, T. The Autonomous Pollination Drone. In Proceedings of the 2nd IFSA Winter Conference on Automation, Robotics & Communications for Industry 4.0 (ARCI' 2022), Andorra, 3 February 2022 ; Volume 1, pp. 38–41.
8. Hashimoto, K. *Visual Servoing*; World Scientific: Singapore, 1993; Volume 7.
9. Doitsidis, L.; Valavanis, K.P.; Tsourveloudis, N.C.; Kontitsis, M. A framework for fuzzy logic based UAV navigation and control. In Proceedings of the IEEE International Conference on Robotics and Automation, New Orleans, LA, USA, 26 April–1 May 2004; Volume 4, pp. 4041–4046.
10. De Schutter, J.; De Laet, T.; Rutgeerts, J.; Decré, W.; Smits, R.; Aertbeliën, E.; Claes, K.; Bruyninckx, H. Constraint-based task specification and estimation for sensor-based robot systems in the presence of geometric uncertainty. *Int. J. Robot. Res.* **2007**, *26*, 433–455. [CrossRef]
11. Strader, J.; Nguyen, J.; Tatsch, C.; Du, Y.; Lassak, K.; Buzzo, B.; Watson, R.; Cerbone, H.; Ohi, N.; Yang, C.; et al. Flower Interaction Subsystem for a Precision Pollination Robot. In Proceedings of the 2019 IEEE/RSJ International Conference on Intelligent Robots and Systems (IROS), Macau, China, 3–8 November 2019; pp. 5534–5541. [CrossRef]
12. Szegedy, C.; Vanhoucke, V.; Ioffe, S.; Shlens, J.; Wojna, Z. Rethinking the inception architecture for computer vision. In Proceedings of the IEEE Conference on Computer Vision and Pattern Recognition, Las Vegas, NV, USA, 27–30 June 2016; pp. 2818–2826.
13. Dias, P.A.; Tabb, A.; Medeiros, H. Apple flower detection using deep convolutional networks. *Comput. Ind.* **2018**, *99*, 17–28. [CrossRef]
14. Guglielmino, P.C.; Seker, Z.; Stallings, N.A.; Craigie, C.A. *Autonomous Drone Pollination*; Worcester Polytechnic Institute (WPI): Worcester, UK, 2021.
15. oi: 10.48550/arxiv.1506.02640 Redmon, J.; Divvala, S.; Girshick, R.; Farhadi, A. You Only Look Once: Unified, Real-Time Object Detection. In Proceedings of the IEEE Conference on Computer Vision and Pattern Recognition (CVPR), Las Vegas, NV, USA, 27–30 June 2016; pp. 779–788. [CrossRef]
16. Howard, A.G.; Zhu, M.; Chen, B.; Kalenichenko, D.; Wang, W.; Weyand, T.; Andreetto, M.; Adam, H. Mobilenets: Efficient convolutional neural networks for mobile vision applications. *arXiv* **2017**, arXiv:1704.04861.
17. Howard, A.; Sandler, M.; Chu, G.; Chen, L.C.; Chen, B.; Tan, M.; Wang, W.; Zhu, Y.; Pang, R.; Vasudevan, V.; et al. Searching for mobilenetv3. In Proceedings of the IEEE/CVF International Conference on Computer Vision, Seoul, Korea, 27 October–2 November 2019; pp. 1314–1324.

18. Ren, S.; He, K.; Girshick, R.; Sun, J. Faster R-CNN: towards real-time object detection with region proposal networks. *IEEE Trans. Pattern Anal. Mach. Intell.* **2016**, *39*, 1137–1149. [CrossRef] [PubMed]
19. He, K.; Gkioxari, G.; Dollár, P.; Girshick, R. Mask r-cnn. In Proceedings of the IEEE International Conference on Computer Vision, Venice, Italy, 22–29 October 2017; pp. 2961–2969.
20. Liu, L.; Ouyang, W.; Wang, X.; Fieguth, P.; Chen, J.; Liu, X.; Pietikäinen, M. Deep learning for generic object detection: A survey. *Int. J. Comput. Vis.* **2020**, *128*, 261–318. [CrossRef]
21. Zhou, B.; Lapedriza, A.; Khosla, A.; Oliva, A.; Torralba, A. Places: A 10 million image database for scene recognition. *IEEE Trans. Pattern Anal. Mach. Intell.* **2017**, *40*, 1452–1464. [CrossRef]

machines

MDPI

Article

Towards Privacy-Preserving Machine Learning for Energy Prediction in Industrial Robotics: Modeling, Evaluation and Integration †

Adam Skuta [1], Philipp Steurer [2], Sebastian Hegenbart [2], Ralph Hoch [2] and Thomas Loruenser [1,*]

[1] AIT Austrian Institute of Technology, 1210 Vienna, Austria
[2] Digital Factory Vorarlberg, 6850 Dornbirn, Austria
* Correspondence: thomas.loruenser@ait.ac.at

† This paper is an extended version of our paper published in Steurer, P.; Skuta, A.; Hegenbart, S.; Hoch, R.; Loruenser, T. Data-Driven Modeling for Privacy-Preserving Energy Prediction of Industrial Robots. In Proceedings of the Automation, Robotics & Communications for Industry 4.0/5.0 2025; Yurish, S., Ed., 2025, pp. 184–189. 5th Winter IFSA Conference on Automation, Robotics & Communications for Industry 4.0/5.0, ARCI' 2025; Conference date: 19 February 2025 Through 21 February 2025.

Abstract

This paper explores the feasibility and implications of developing a privacy-preserving, data-driven cloud service for predicting the energy consumption of industrial robots. Using machine learning, we evaluated three neural network architectures—dense, LSTM, and convolutional–LSTM hybrids—to model energy usage based on robot trajectory data. Our results show that models incorporating manually engineered features (angles, velocities, and accelerations) significantly improve prediction accuracy. To ensure secure collaboration in industrial environments where data confidentiality is critical, we integrate privacy-preserving machine learning (ppML) techniques based on secure multi-party computation (SMPC). This allows energy inference to be performed without exposing proprietary model weights or confidential input trajectories. We analyze the performance impact of SMPC on different network types and evaluate two optimization strategies, using public model weights through permutation and evaluating activation functions in plaintext, to reduce inference overhead. The results highlight that network architecture plays a larger role in encrypted inference efficiency than feature dimensionality, with dense networks being the most SMPC-efficient. In addition to model development, we identify and discuss specific stages in the MLOps workflow—particularly model serving and monitoring—that require adaptation to support ppML. These insights are useful for integrating ppML into modern machine learning pipelines.

Keywords: machine learning; energy prediction; privacy-preserving machine learning

1. Introduction

The growing emphasis on energy efficiency in production processes is driven by increasing operational costs and stringent environmental regulations. Industrial production systems consume a significant amount of energy, making energy modeling a critical tool for optimizing resource utilization [1]. Among production systems, industrial robots are a key feature due to their widespread adoption and energy demands in automated workflows [2]. Modeling such systems, especially with the use of deep learning, might require extensive amounts of data and computational resources. These resources can often be accessed through modern cloud infrastructure and frequently involve sharing sensitive operational data during collaboration. In collaborative industrial environments, privacy-preserving machine learning (ppML) is emerging as a key methodology to ensure secure collaboration on data gathering while protecting intellectual property across diverse applications.In this work we aim to develop a comprehensive understanding and framework for integrating data-driven energy modeling with ppML in a seamless way along the development life cycle in order to enable easy development of novel collaborative cloud-based use cases for smart manufacturing. This paper is an extended version of our paper published in [3].

1.1. Motivation

Energy models offer significant value, but their deployment can involve sharing sensitive intellectual property (e.g., neural network weights or trajectory data). This poses challenges in collaborative environments where multiple stakeholders—such as manufacturers and third-party service providers—must interact without exposing proprietary information. Generally speaking, there has been a significant surge in machine learning adoption in the industrial sector, with applications ranging from predictive maintenance to process optimization. As these ML models grow increasingly sophisticated, their computational demands have led to a paradigm shift toward cloud-based solutions typically managed by third-party service providers. This transition, while offering substantial computational benefits, introduces notable privacy and security concerns for industrial stakeholders. Privacy-preserving machine learning (ppML) is a promising solution to address these challenges. ppML encompasses techniques and methodologies designed to protect both the confidentiality of training data and the privacy of model parameters during the training and inference phases. The primary objective is to prevent unauthorized access to sensitive information while maintaining the functionality and performance of ML models.

The motivations for this work are twofold. One the one hand, based on preliminary findings [3], we want to demonstrate the capabilities of ppML in robotics. Through extensive benchmarking of different models to solve a particular use case, we explore the practicability of SMPC for industrial applications in general. The results should help developers of industrial cloud-based services and platforms to estimate the overhead of end-to-end security and provide ways to optimize performance. On the other hand, we want to deliver a blueprint for the development of privacy-preserving applications themselves, such as the one visualized in Figure 1, thus streamlining the software development and ppML integration processes for application developers. Only if the development process is integrated into the design methodology will it be feasible for non-experts in secure computation techniques to leverage the potential of ppML and thus design novel collaborative use cases based on data for future production systems.

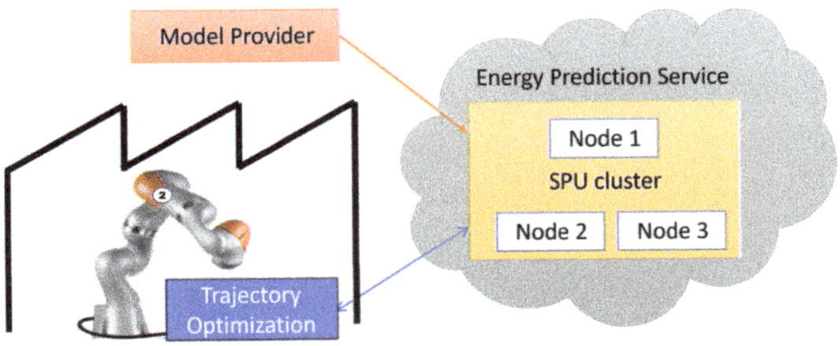

Figure 1. High-level overview of a cloud-based energy prediction use case integrating a privacy-preserving SMPC workflow [3].

1.2. Related Work

In the following section we discuss the state of the art in the two main fields relevant for our work, i.e., the work that has been conducted on data-driven energy modeling in manufacturing and the field of privacy-preserving computing for machine learning.

1.2.1. Energy Modeling

For modeling the energy consumption of industrial processes, several approaches have been utilized including (i) analytical modeling where energy demand is modeled by analyzing different operation modes (e.g. on, off, idle, etc.) and defining the energy consumption of each mode, usually represented as an average demand; (ii) empirical modeling, which commonly uses multiple linear regression to fit a predefined mathematical expression that is linear in the models' parameters; (iii) physical modeling where fundamental physical relationships are expressed in mathematical equations; (iv) machine learning methods, which utilize data-driven algorithms to identify relationships between input and output data [4]. For analytical modeling, oversimplification can be considered as a possible drawback, while for empirical modeling it can be difficult to find the right mathematical expression that fits the collected data [4]. Physics-based models can be very accurate, but they are often difficult to implement and require a high level of insight and expertise about the problem to be modeled [1,4]. Data-driven approaches such as neural networks offer a flexible alternative that harnesses network approximation capabilities, although with acknowledged tradeoffs in terms of accuracy and extrapolation compared to, e.g., physics-based models.

Concerning energy consumption modeling for industrial robots, research was conducted based on the development of a physical model [5–7], which commonly requires the identification of parameters for robot dynamics and parameters describing power losses due to, e.g., friction, copper losses in the motors or losses in the control system. For dynamic parameter identification torque measurements must usually be obtained [8–10]. However, for some industrial robots, it is not trivial to determine the torque of the robot axis, which is addressed in [5], where the dynamic robot parameters are identified based only on the total electrical power consumption. In [11], modeling is performed using the Modelica language, while dynamic parameters are obtained from documentation and related studies. Works applying data-driven approaches to the robotic energy modeling problem include strategies such as the utilization of neural networks for determining operation parameters and optimizing energy consumption. Zhang and Yang [12] use a feedforward neural network to determine the operating parameters of maximum velocity and acceleration with a focus on minimizing total energy consumption on a given path and type of motion. Yin [13] et al. used a deep neural network for modeling and optimizing the energy consumption of a robot on a simplified case study. In [1], different states of operation are identified which are the approximated using linear regression.

In general, energy modeling using machine learning techniques has provided promising results and is increasingly applied in industrial manufacturing processes. However, regarding energy modeling for IRs, more research needs to be conducted in terms of case studies on real applications and by applying different machine learning algorithms in order to compare their performance.

1.2.2. Privacy-Preserving Machine Learning

The concept of SMPC was invented more than 30 years ago, but for a long time, it was considered only of theoretical interest. However, progress in recent years has led to many interesting applications which can be realized with practical efficiency (e.g., as in [14,15]). Some SMPC frameworks with general use-cases like MPyC (https://github.

com/lschoe/mpyc (accessed on 7 January 2025)) support simple neural networks or MP-SPDZ [16], which enables simple neural networks or the importing of neural networks from well-known frameworks like PyTorch 2.8. This progress, along with the popularity of deep-neural network usage, has led to the development of many SMPC frameworks developed specifically for neural network usage.

SecretFlow-SPU [17] is a multi-party computation (MPC) framework suitable for machine learning that currently supports three MPC protocols, namely ABY3 [18], Semi2k-SPDZ [19], and Cheetah [20]. ABY3 [18] is a framework for ppML across three servers using secure three-party computation. The framework enables efficient conversion between arithmetic, binary, and Yao computation methods, incorporating new techniques for fixed-point multiplication and polynomial time evaluation. CrypTen [21] is an MPC-enabling framework for machine learning. It utilizes a semi-honest threat model and abstracts many MPC primitives in the form of tensor computation, automatic differentiation, and many widely used neural network layers.

1.3. Contribution and Structure

This paper significantly extends preliminary work from [3]. Specifically, the contributions are, first, a comprehensive research and evaluation of data-driven energy prediction for industrial robots with high accuracy based on small models for later ppML integration; second, a comprehensive analysis of ppML integration for the given models specifically considering performance optimization and privacy–security trade-offs; and third, an analysis and discussion of ppML integration into MLOps workflow to support seamless development for non-security experts in support of future applications.

To achieve this, we first develop a neural network-based prediction model for energy consumption. We then apply ppML methods to these models to achieve data safety and integrity in collaborative environments. Finally, we investigate and locate the parts of the MLOps pipeline that would require attention specific to ppML integration.

The remainder of the paper is structured as follows: In Section 2 the process of acquiring the data for training and testing from the industrial robot is explained, followed by Section 3 where feature engineering, data preprocessing, and different model architectures developed are shown. In Section 4 the framework for privacy-preserving energy prediction is introduced along with a brief explanation of the two optimization methods used for decreasing inference times in the ppML domain. Next, in Section 5 an analysis of the developed model's results is conducted in both plaintext and ppML forms, followed by Section 6, where MLOps is introduced, and its required changes for ppML integration are discussed. Section 7 concludes this paper.

2. Data Acquisition

In this section a developed method for robot data acquisition is presented in detail, as dedicated hardware and software were necessary to acquire data with high accuracy and quality.

2.1. Boundary Conditions

The system for which energy consumption is to be modeled is the seven-axis collaborative industrial robot LBR iiwa R800, developed by KUKA Germany GmbH, Augsburg, Germany. The main components of the robotic system are depicted in Figure 2a. The robot is integrated into a laboratory setup that simulates a production line, where it is employed for assembly and handling tasks. Robot applications and motion programming are developed using the KUKA Sunrise.Workbenchdevelopment environment. Within this environment, the KUKA RoboticsAPI is available, offering functionalities for motion planning, execution, and data recording.

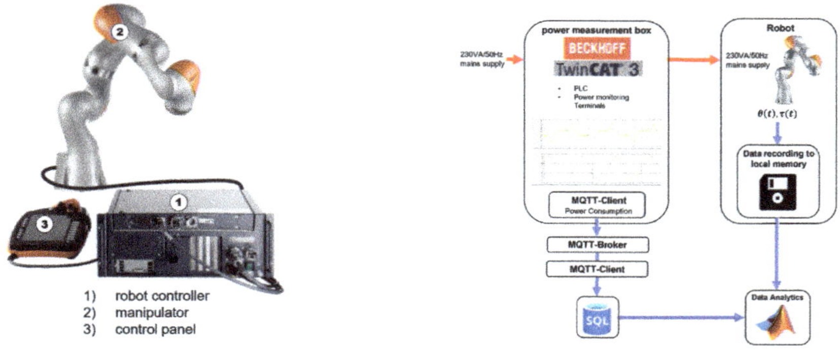

Figure 2. Data acquisition model components and structure [3]. (**a**) Components of the robotic system (adapted from [22]); (**b**) general structure of the measurement concept.

2.2. Data Acquisition

To enable the development of data-driven models, it is essential to first collect meaningful data that captures relevant aspects of the system's behavior. For the problem addressed in this work, the relevant data includes

- Robot axis angles over time, $\theta(t)$
- Consumed electrical active power, $p(t)$

The measurement concept designed to obtain data from both the electrical and mechanical domains is illustrated in Figure 2b. The figure outlines the general structure and operational principle of the data acquisition setup. The flow of electrical energy is represented by red arrows moving from left to right. As shown, the robot's electrical supply passes through an additional power measurement box, which measures the robot's power consumption. For power measurement, voltage $v(t)$ and current $i(t)$ are sampled at a rate of 20,000 $\frac{samples}{second}$ and aggregated over two mains supply periods, corresponding to $\Delta T = 40$ ms. These aggregated values are then used to compute the active power. Mechanical data is captured using functions provided by the KUKA RoboticsAPI, which logs motion data directly to a file. From the mechanical domain, axis angles $\theta(t)$ and measured axis torques $\tau(t)$ are recorded at a sampling interval of 40 ms. The mechanical data flow is indicated by blue arrows. Electrical data is transmitted using the MQTT network protocol and stored in a central SQL database. Both the data collected from the database and from the robot controller are subsequently available for further analysis.

2.3. Trajectory Generation and Execution

Sufficient excitation is necessary to obtain enough data for the creation of accurate models by machine learning. To this end, we use two different methods to generate trajectories suitable for excitation, namely

- Limited Space Random (LimSp)—where random positions within a defined subspace of all reachable manipulator poses are generated. The orientation of the end-effector was fixed.
- All Space Random (AllSp)—where random robot configurations, i.e., axis angles, are generated using the entire configuration space of the robot. Valid robot configurations are subject to physical constraints, e.g., collisions with environment.

Limited Space Random has the advantage that trajectories can be generated more easily and that models, due to reduced complexity, can faster mimic system behavior. However, such models are likely to not perform well outside the provided training data by extrapolation. To develop models that generalize well over all possible robot movements and to test the limitations of the models obtained by Limited Space data, the All Space data is generated as well.

For both methods, we created several trajectories (46 Lim.Sp., 30 AllSp.) and stored their robot configuration, which can be later used as inputs for execution on the real robot. Trajectories were executed as ptp motions between given positions. To get additional variability and to produce more data out of the same source trajectories, each motion is executed with a random blending factor and random velocity. Figure 3 shows an example of one LimSp. and one AllSp. trajectory, with the effect of blending also shown. For Limited Space data 202 recordings corresponding to 2.8 h recording time were made; for All Space 135 recordings and 2.7 h.

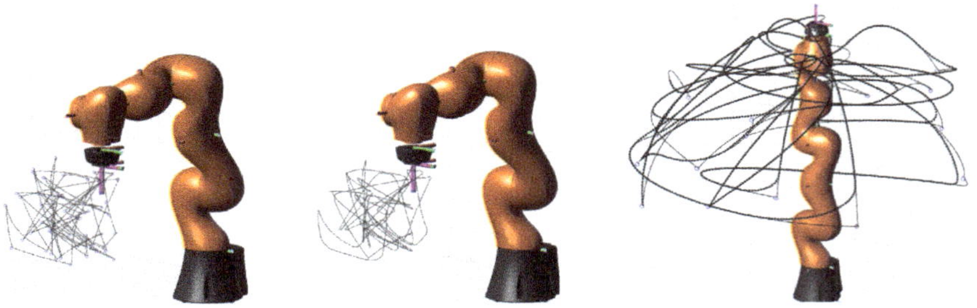

(a) (b) (c)

Figure 3. Effect of relative blending factor: (**a**)—example of Limited Space Random points indicated with circles and the ptp motion between points as a solid line; (**b**)—recordings when trajectory (**a**) is executed with blending; (**c**)—example of All Space Random points indicated as circles, ptp motion as a solid line, and recorded motion as a dotted fat line. Taken from [3].

2.4. Data Preparation and Analysis

The data was recorded by two different systems: the electrical data from the measurement box and the mechanical data from the robot controller. Each of these systems operates independently, resulting in both systems generating their own timestamps, so the measurements must be combined using an appropriate method. To this end, the obtained time series are combined with a common time vector with sampling interval 40 ms using linear interpolation between points.

In addition to merging of the data, we also examined how the time series fit together by observing distinctive patterns in the data. For comparison of the electrical and mechanical data, the mechanical power p_{mech} is calculated with the angular velocity $\dot{\theta}$ and measured axis torque τ (1).

$$p_{mech} = \tau \cdot \dot{\theta} \tag{1}$$

Angular velocity is calculated out of the measured axis angles by calculating the first-order centered finite difference (2).

$$\dot{\theta}(t) \approx \frac{\theta(t+\Delta T) - \theta(t-\Delta T)}{2\Delta T} \tag{2}$$

Figure 4 top shows the resulting mechanical and electrical power of a selected measurement. Apart from the initial peak, attributed to inrush currents, the results are similar in amplitude and course, but a considerable shift in time can be observed.

This time difference in the acquired data can be attributed to non-synchronized clocks of the measurement systems. To compensate for this overall offset, or lag, the cross-correlation of the two time series was calculated, and the time difference to compensate for was determined as the lag at which cross-correlation becomes maximal. Figure 4 shows an example of the original and the shifted data. The sequences are shown in the time domain along with a scatter plot for visualizing the correlation.

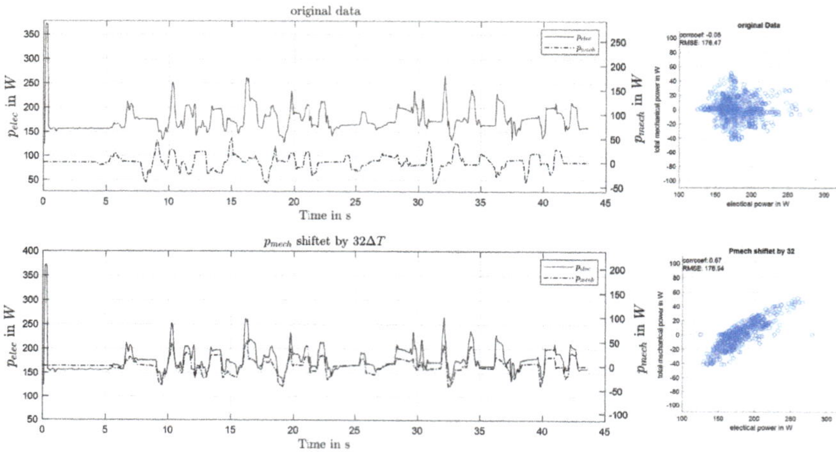

Figure 4. Comparison of shifted and non-shifted power sequences.

3. Model Development

In the following section the development of energy models is assessed for our case studies. The used modeling approaches are based on machine learning techniques as well as neural networks. Multiple network architectures are explored, including long short-term memory (LSTM) networks with engineered features as inputs, LSTM networks using a convolutional head for automated feature extraction, and dense networks. Specifically, we train neural networks based on certain input features derived from robot trajectory data to predict the energy consumption in a sequence-to-sequence-based problem.

3.1. Feature Engineering

To select suitable features, we will first take a closer look at the system shown in Figure 5. Our input to the system is the desired trajectory $\theta_{ref}(t)$, and the system response we are interested in is the electrical power supplied $p_{elec}(t)$. The desired trajectory goes as reference input to the robot controller, which ensures that the angles $\theta(t)$ of the manipulator axis follow the reference as closely as possible.

Portions of the total power supplied are lost due to losses in the control system, losses in the motors, as well as losses through friction. What remains is the mechanical power that can be used for performing various tasks.

As already shown in Section 2.4, the mechanical power is closely related to the electrical power and can be calculated by ($p_{mech} = \tau \cdot \dot{\theta}$). In robotics, the torque acting on a robot axis can be calculated by solving the so-called inverse dynamics problem, which is given as a set of nonlinear differential equations in the variables angle θ, angular velocity $\dot{\theta}$, and angular acceleration $\ddot{\theta}$ [23].

$$\tau = M(\theta)\ddot{\theta} + h(\theta,\dot{\theta}) \tag{3}$$

where $M(\theta)$ is a symmetric, positively defined mass matrix, and $h(\theta,\dot{\theta})$ are forces that combine centripetal, coriolis, gravitational, and frictional terms. By multiplying (3) with $\dot{\theta}$, the mechanical power of the system under consideration is given by a nonlinear function with the variables angle θ, angular velocity $\dot{\theta}$, and angular acceleration $\ddot{\theta}$. Since the electrical power is closely related to the mechanical power, and the mechanical power depends substantially on the three variables: θ, $\dot{\theta}$, and $\ddot{\theta}$. These variables can be considered as suitable features to model the system behavior via machine learning algorithms.

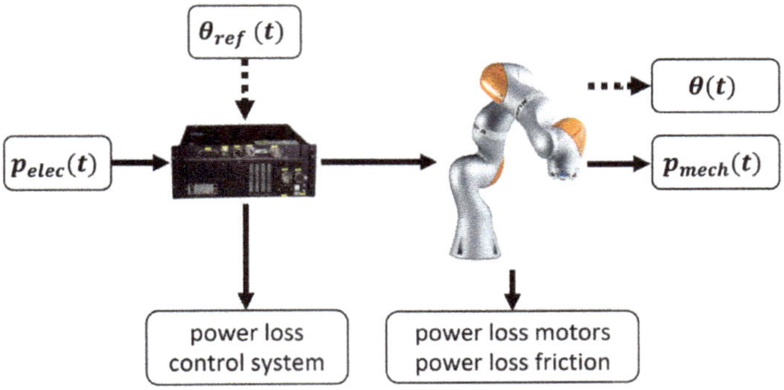

Figure 5. Energy flow through the robot system [3]. The target variable for modeling is the electrical power supplied $p_{\text{elec}}(t)$.

For extraction of the engineered features, the first- and second-order finite difference method is applied to the recorded data ((2) and (4)).

$$\ddot{\theta}(t) \approx \frac{\theta(t+\Delta T) - 2\theta(t) + \theta(t-\Delta T)}{(\Delta T)^2} \tag{4}$$

Before the collected data is handed over to the selected modeling algorithms, normalization is performed. For the electrical power, normalization is chosen with $p_{norm} = (p_{elec} - p_{offset})/p_{scale}$, with power normalization parameters are chosen with $p_{offset} = 155$ and $p_{scale} = 166$.Angle, angular velocity, and angular acceleration are scaled by the factors $\theta_{scale} = 3$, $\dot{\theta}_{scale} = 1.6$, and $\ddot{\theta}_{scale} = 13$.

3.2. Models

Three different distinct model architectures were developed, namely dense models, LSTM models, and LSTM models with convolutional layers, as can be seen in Figure 6. Both the dense and LSTM models can take a variable amount of input features. One network has only the normalized seven axis angles as input, the second additionally has the normalized angular velocities as input, which leads to a 14-dimensional input, and the third network has all 21 constructed features as input.

The dense models consist of four linear layers with a ReLU activation function. It accepts features in the format (b, f), where b is the batch size, and f is the number of features for one timestamp. The LSTM network structure consists of a sequence input layer, followed by an LSTM layer with 10 hidden units, succeeded by a four-layered dense network without activation functions. The last layer is a one-dimensional output layer. It accepts features in the format (b, t, f), where b is the batch size, t is the timestamps, and

f is the individual features. With this structure, three networks were set up that differ only in their input dimension. Importantly, the datasets used for LSTM and dense models are identical, only the input arrays are reshaped to match the required format of each architecture before training and evaluation.

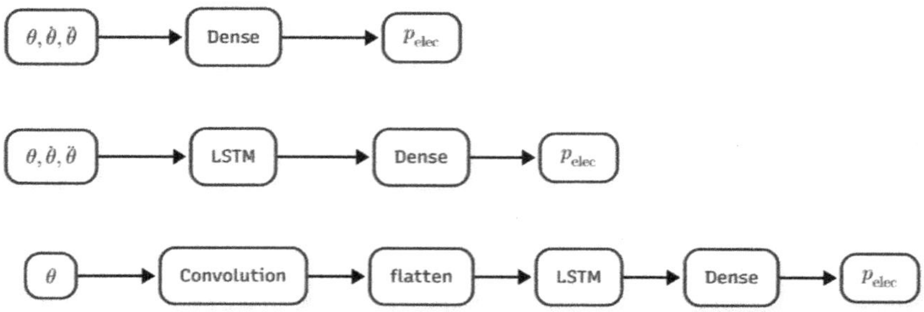

Figure 6. Architectures of the models developed ranging from the least to the most complex. From top to bottom: dense architecture, LSTM architecture, and LSTM with convolutional layer architecture.

LSTM with convolutional layer for feature extraction is a model structure that is capable of automatically extracting features from the data without requiring any prior knowledge of the system. It is interesting to note that the derivative of a signal can be computed by convolution of the signal with an appropriate filter kernel. Common filter kernels are therefore the first- and second-order derivative of a Gaussian g(t) ((5) and (6))[24].

$$\frac{d}{dt}f(t) \approx f(t) * \left(\frac{d}{dt}g(t)\right) \tag{5}$$

$$\frac{d^2}{dt^2}f(t) \approx f(t) * \left(\frac{d^2}{dt^2}g(t)\right) \tag{6}$$

When the width of the kernels is reduced to L = 3, then the convolution of the resulting kernels with a signal is the same as the calculation of the first and second finite differences given by (2) and (4) for $\Delta T = 1$. Hence, a convolutional layer with multiple filter kernels of width 3 prior to the LSTM layer can, in principle, provide the same performance as the networks with manually extracted features. For this architecture, the input features are arranged in the format (b, L, f), where b is batch size, f is the individual features, and L corresponds to the kernel width. The datasets remain the same but are reshaped so that the CNN consistently receives L timestamps, with appropriate padding applied at the sequence boundaries.

The tested network structures are LSTM, dense, and LSTM with preceding convolutional layers. Shown in Table 1 are the tested structures and their number of parameters for complexity comparison.

We use LSTM (lstm-) and dense (mlp-) networks with different input features. We either use only the angular values θ (-Nf7), angular values and angular velocity $\theta, \dot{\theta}$ (-Nf14), or all $\theta, \dot{\theta}, \ddot{\theta}$ (-Nf21) as input to the neural nets. For dense versions, small architectures (-small) with similar numbers of parameters as compared to the LSTMs as well as architectures with higher parameter numbers (-large) were assessed. For LSTM with convolutional layer (cnn-), we trained two networks with three kernels (-k3) and a varying kernel width of $L = 3$ (-L3) and $L = 5$ (-L5). This configuration only expects the angular values θ. For model training and evaluation, the datasets were split into training, testing, and validation sets. When splitting, we ensured that no recordings based on the same source trajectory were found in the same split. All models were trained using Adam optimizer [25].

Table 1. Number of parameters for each model [3].

Name	Number of Parameters
lstmNf7	903
lstmNf14	1183
lstmNf21	1463
mlpNf7-small	910
mlpNf14-small	1192
mlpNf21-small	1477
mlpNf7-large	2625
mlpNf14-large	3073
mlpNf21-large	3521
cnn-k3L3	1475
cnn-k3L5	1481

4. Privacy-Preserving Collaborative Energy Prediction

To address the challenges related to disclosing sensitive information, we introduce a ppML framework that utilizes SMPC techniques for energy modeling. This framework ensures that

- Model providers maintain the confidentiality of their trained energy models.
- Clients can perform encrypted inference on proprietary trajectories without exposing sensitive data.

SMPC can be computationally intensive. In scenarios where real-time or near real-time performance is required—such as energy optimization—slow computation can be a limiting factor for certain approaches. It is important to note that the computations performed by the adopted methodology are accurate up to floating-point rounding errors. As a result, the prediction accuracy of the underlying networks remains unaffected.

To ensure that the performance requirements of our use case are satisfied, we evaluate three state-of-the-art secure multi-party computation (SMPC) frameworks: MP-SPDZ [16], CrypTen [21], and SecretFlow-SPU [17]. For improved reproducibility, two well-established and widely adopted neural network architectures—AlexNet [26] and LeNet [27]—were used. Inference was conducted using SMPC following a clear-text training phase. It is important to note that training neural network models in the encrypted domain was not feasible due to performance limitations. As shown in Table 2, SecretFlow-SPU achieved the lowest inference time per sample. Combined with its user-friendly interface and ease of implementation, SecretFlow-SPU was chosen as the primary framework for this work.

Table 2. Comparison of the inference times of several SMPC frameworks on two open-source models. The times shown are for single sample inference [3].

Framework	Model	Time [s]
MP-SPDZ	LeNet	0.15
MP-SPDZ	AlexNet	1.08
CrypTen	LeNet	0.42
CrypTen	AlexNet	3.84
SPU	LeNet	0.08
SPU	AlexNet	0.34

4.1. Overview of SPU

For privacy-preserving inference, the SecretFlow-SPU framework [17] was employed. The architecture of the SPU framework is illustrated in Figure 7. Unlike other frameworks such as CrypTen [21] and MP-SPDZ [16], SPU can integrate machine learning models from any popular framework that can be compiled to the higher level operations (HLO) intermediate representation, requiring only minimal modifications to the training and inference scripts. The SPU frontend processes this HLO representation and transforms it into a SPU-specific PPHLO representation, where additional MPC-specific optimizations and data visibility annotations are applied. This representation is then passed to the SPU backend, which operates across multiple nodes based on the underlying MPC protocol. SPU follows the single-program-multiple-data (SPMD) model, meaning all nodes receive and execute the same code.

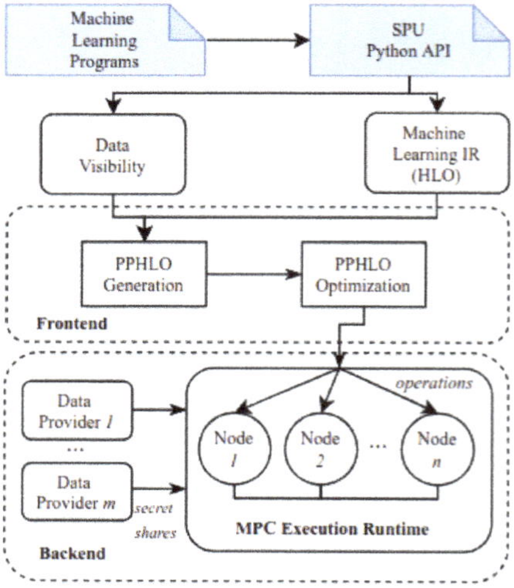

Figure 7. Overview of SPU architecture [17]. For machine learning development the main interaction is with SPU Python API.

4.2. Problems ppML

The major obstacle that ppML using SMPC presents is the increased computational and communication overhead required for the computing nodes. This is illustrated in Table 3, where a noticeable difference between inference of plaintext models and SMPC models can be observed. Coupled with other real-world factors like network latency and packet loss, a faster inference speed for SMPC is required in order to achieve real-world adoption. Several optimization strategies have been proposed and evaluated in the past [28–30]. However, these are mostly used for large language models (LLMs), and their effect on smaller networks similar to the ones we have developed is untested. Two such optimizations are evaluated in this paper, namely enabling plaintext weights through model permutation and evaluating activation functions in plaintext.

Table 3. Comparison of plaintext vs. SMPC inference times.

Model	Time Plaintext [s]	Time SMPC [s]
LeNet	0.01	0.08
AlexNet	0.26	0.34

Typically, SPU or other popular SMPC frameworks view neural network inference as a function of two inputs, those being the input features (robots angles and their derivations in this case) and trained model weights. A typical inference in SPU secret shares both of those inputs, making them illegible for honest-but-curious adversaries. While this approach is secure for both inputs, it suffers from larger computational overhead and therefore larger inference times, making it harder to apply for real-life applications. One way to speed up the inference is to make one of the inputs public. In the case of open-source models, the biggest candidates are the pre-trained model parameters. To still ensure a secure model inference, linear layer parameters and input features can be permuted using predetermined random permutation matrices [28]. This way linear layers can be kept public, and the input features will be used in their permuted secret-shared state. This has the potential to drastically reduce inference times for fully connected linear layers without sacrificing too much of the security.

Another way of possibly reducing model inference times in SMPC is to evaluate the activation functions in plaintext. Activation functions in SMPC inference take, contrary to the plaintext inference, the majority of the inference time [29]. Different activation functions have different impacts on inference times and there has been an effort to minimize this added overhead by simplifying the activation functions. In the case of our developed networks, ReLU activation is used. This belongs to the class of already very simple functions, and further reducing it would introduce a significant error to the utility of the network. Therefore, a different approach was chosen to investigate, which is the evaluation of activations in plaintext. The secret-shared intermediary values can be sent to and reconstructed by the data owner, who then performs the activation locally in plaintext and again secret-shares the resulting values to the nodes. This process is repeated until all activations in the network have been evaluated.

5. Evaluation Results

In this section, all our developed models are evaluated. First, the qualitative results of the models are discussed purely in plaintext form. Second, the inference times of the models are discussed in SMPC form. As the quality of the models does not change compared to plaintext, only inference time is taken into consideration as the main metric. Third, the impact of the two proposed optimization strategies is discussed on the best-performing model. Finally, the deployment considerations for SMPC models are evaluated, and the impact of network delay and packet loss is analyzed.

5.1. Models' Quality Evaluation

The trained models are evaluated using the metrics mean absolute error (MAE), root mean squared error (RMSE). and mean absolute percentage error (MAPE) given the following equations:

$$MAE = \frac{1}{N}\sum_{k=1}^{N}|y_k - \hat{y_k}| \tag{7}$$

$$RMSE = \sqrt{\frac{1}{N}\sum_{k=1}^{N}(y_k - \hat{y_k})^2} \tag{8}$$

$$MAPE = \frac{100}{N}\sum_{k=1}^{N}\frac{|y_k - \hat{y_k}|}{|y_k|} \tag{9}$$

Metrics computed from the combined LimSp and AllSp datasets are shown in Table 4. The results indicate that the manually engineered features are informative. Using the complete set of θ, $\dot{\theta}$, and $\ddot{\theta}$ features (-Nf21) resulted in the lowest error rates. In contrast, models without CNN layers and trained solely on seven axis angles (-Nf7) showed the weakest performance. Although the MAE for convolutional models is slightly higher compared to models using all three input feature types (-Nf21), the difference remains within a comparable range. Notably, LSTM networks outperform fully connected models when only θ or $[\theta, \dot{\theta}]$ are used as inputs. However, when all features are included, fully connected models achieve better performance. Furthermore, the number of weights in the dense models does not appear to have a significant impact on performance, as the metrics remain stable across different configurations.

Table 4. Comparison of metrics for tested model structures. Values are calculated from the combined LimSp and AllSp datasets [3].

Name	MAE [W]	RMSE [W]	MAPE [%]
mlpNf21-large	3.279	5.127	1.800
mlpNf21-small	3.371	5.180	1.839
lstmNf21	3.558	5.548	1.943
cnn-k3L5	3.595	5.719	1.958
lstmNf14	3.676	6.659	2.033
cnn-k3L3	4.341	7.417	2.299
mlpNf14-small	4.559	7.145	2.496
mlpNf14-large	5.446	9.427	2.950
lstmNf7	10.041	13.850	5.540
mlpNf7-small	15.269	20.172	8.430
mlpNf7-large	16.368	21.343	9.111

To assess the impact of different datasets on model performance, metrics were computed separately for the LimSp and AllSp datasets. Table 5 summarizes the results, with mean absolute error (MAE) used as the primary evaluation metric. The results show that performance is notably better on the LimSp dataset. This outcome is expected, as the AllSp dataset covers a much larger configuration space, which in turn requires a substantially greater number of samples for comparable performance.

Table 5. Comparison of mean absolute error for tested model structures. Values are calculated from the combined LimSp and AllSp datasets [3].

Name	LimSp	AllSp	Lim + AllSp
mlpNf21-large	2.043	5.270	3.279
mlpNf21-small	2.230	5.090	3.371
lstmNf21	2.288	5.604	3.558
cnn-k3L5	2.226	5.800	3.595
lstmNf14	2.431	5.682	3.676
cnn-k3L3	2.735	6.929	4.341
mlpNf14-small	3.911	5.603	4.559
mlpNf14-large	3.867	7.991	5.446
lstmNf7	7.999	13.333	10.041
mlpNf7-small	14.148	17.074	15.269
mlpNf7-large	14.355	19.612	16.368

The difference in performance is depicted in Figure 8, which shows predictions for selected sequences from selected model mlpNf21-large. The sequences illustrate the worst predictions according to the MAE value for the LimSp and AllSp datasets, respectively.

The results indicate that the model performs effectively on the LimSp dataset, with predictions closely matching the measured values—even for the lowest-performing sequences. In contrast, the AllSp dataset reveals discrepancies in certain parts of the sequences, suggesting that model performance may be limited by the relatively small number of available training samples. Overall, these findings support the feasibility of using a data-driven approach with neural networks for energy prediction modeling.

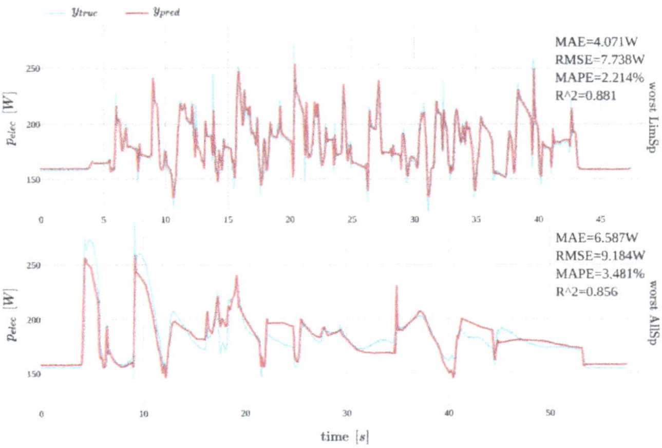

Figure 8. Comparison of the worst predictions on LimSp and AllSp datasets with respect to MAE using model mlpNf21-large.

5.2. Models Evaluation Under SMPC

The trained models are evaluated using inference time under SMPC and inference time in plaintext form. All the benchmarks were established on an Intel Xeon W-2245 CPU, 126 GB of RAM using Ubuntu 24.04 and Python 3.10.14. The SMPC protocol chosen was ABY3 with three computing nodes and two provider nodes, one providing the inputs and the other one providing the trained model weights. All tests were conducted within a local area network. Table 6 presents the performance measurements for SecretFlow-SPU using the developed energy prediction networks. The results highlight the following observations:

- Feature dimensionality has minimal impact on inference performance in the encrypted domain.
- The type of network layers (recurrent, convolutional, or dense) plays the most significant role in determining performance when used with SPU.

Convolutional networks exhibit increased computational demands in the encrypted domain. In contrast, dense architectures, which require fewer operations, are particularly well-suited for SMPC environments. Recurrent networks such as LSTMs, while composed of relatively simple operations, are less amenable to parallelization due to their sequential computation of hidden and cell states.

Table 6. Comparison of inference time for all models in both SMPC and plaintext form. For SMPC inference both model weights and inputs are secret shared [3].

Model	SMPC [s]	Plaintext [s]
lstmNf7	0.561	0.180
lstmNf14	0.562	0.175
lstmNf21	0.562	0.177
cnn-k3L5	0.573	0.183
cnn-k3L3	0.573	0.183
mlpNf7-large	0.154	0.025
mlpNf14-large	0.154	0.025
mlpNf21-large	0.154	0.025

5.3. *Analyzing Performance-Privacy Trade-Off*

The impact of the two optimization strategies mentioned in the previous chapter on the encrypted inference time was studied as well. The results can be seen on Table 7. The model chosen is the MLP 21 due to its best performance in both inference time and qualitative results and due to its simple architecture, which made experimentation easier.

Table 7. Impact of different optimization attempts on mlpNf21-large. Time encrypted shows how the optimization affects the SMPC inference. For permutation of weights this shows the impact of keeping the model weights public and permuted, while the inputs are permuted and secret-shared.

Optimization Used	Time Encrypted [s]	Time Plaintext [s]
Activation functions in plaintext	0.156	0.024
Permutation of weights	0.065	0.022

For the first strategy of evaluating activation functions in plaintext form, slightly worse inference time was measured. This could be because the time required for collecting shares, reconstructing them, and then again resharing the outputs multiple times during forward pass is more computationally expensive than just a simple evaluation of the function in SMPC form. This optimization can prove useful for more complex neural networks with more complicated and less frequent activations and is subject to further study.

The second strategy of permuting both the weights and inputs, while keeping the weights in plaintext form, proved to be a significant speed increase resulting in a 2.4-times inference time decrease compared to a full encrypted MLP 21 inference. This is an expected outcome, as computing on one less encrypted output significantly reduces the computational overhead.

5.4. *Deployment Considerations*

The impact of network delay and packet loss on the inference has also been studied, as these metrics are an important factor to consider [31], especially when talking about SMPC inference, as it relies heavily on inter-node communication for some operations. As can be seen on Figure 9, the effect of network delay linearly increases the inference time for all models, but the LSTM and LSTM with convolution have much steeper slopes and are therefore more affected. From the figure, it can also be seen that both LSTM and LSTM with convolution have almost the same increase, which hints at the fact that the LSTM layer affects this metric the most. Similarly for packet loss, the relationship seems to be almost linear, with some visible nonlinearities occurring with higher packet loss, especially for models using LSTM layers. However, scenarios involving very high packet loss are not the primary focus of this work, as they reflect severe communication issues rather than typical real-world conditions.

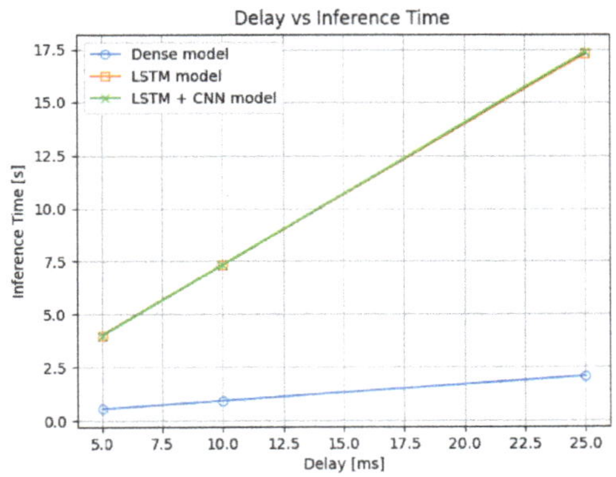

(a)

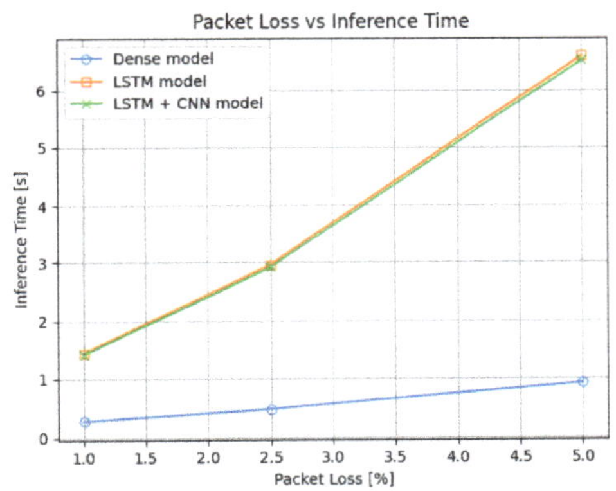

(b)

Figure 9. Effects of network complications on inference time: (**a**) effect of added network delay; (**b**) effect of added packet loss. Note that both models using LSTM layer have almost identical inference times under these conditions and are therefore overlapping in the graphs.

6. Integration into MLOps

Machine learning operations (MLOps) are a set of best practices for experimenting, developing, deploying, and monitoring of machine learning (ML) systems in an end-to-end manner. Bringing the ML system to production involves several steps shown in Figure 10.

The amount of automation of these steps dictates the level of maturity of the ML system (https://cloud.google.com/architecture/mlops-continuous-delivery-and-automation-pipelines-in-machine-learning (accessed on 30 June 2025)). When it comes to impact of privacy-preserving machine learning inference on these steps, especially the final two steps, model serving and model monitoring are most affected.

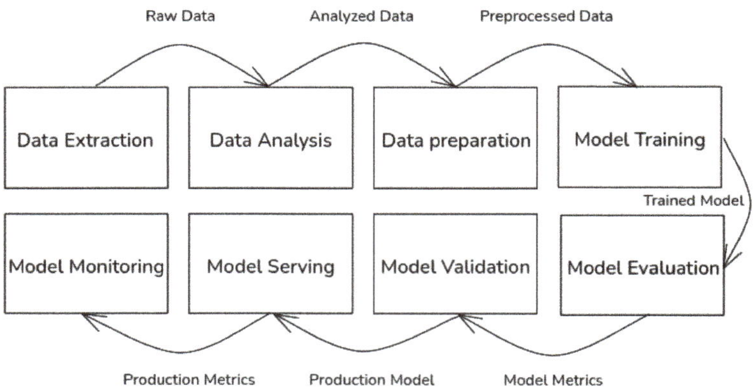

Figure 10. Data science steps required for the development of a machine learning system.

6.1. MLOps Workflow

An example of an MLOps workflow can be seen in Figure 11. One of the key differences between MLOps workflows and non-MLOps workflows is the development target. In non-MLOps workflows, a trained and validated model is deployed to production and served for inference. In MLOps workflows an automated pipeline that conducts the data processing, model training, and evaluation is deployed to production. This pipeline, if the resulting model is deemed ready for production, pushes the model to a model registry, which serves as a centralized repository for model structure and artifacts. From here the model-serving service takes the latest, best model and conducts prediction service. The prediction service is then monitored and acts as a trigger to initiate the deployed pipeline again if the model's performance degrades under a certain threshold.

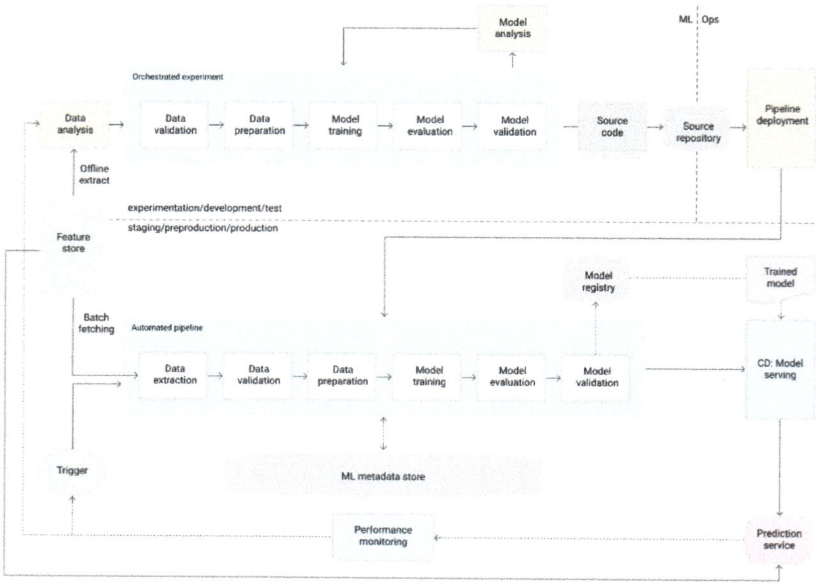

Figure 11. Example of an MLOps workflow. Taken from [32].

6.2. Towards a ppMLOps Workflow

To integrate ppML into the existing approach, the following extensions would be required, as encountered in our experiments:

Model serving is a process where the validated model that is being promoted to production is being served to be used for inference by the end user. Services that provide this process are typically cloud providers (Amazon AWS, Microsoft Azure, Google Cloud, etc.). In the context of ppML, the inference process generally involves three parties:

- The model owner, who provides the model structure as well as the trained parameters of the latest validated model,
- The end user, who provides private data for inference, and
- The SMPC inference service, which processes inputs from the other two parties and returns the result to the End User.

Among these, the SMPC service bears the majority of the complexity. The developers of this service should abstract the SMPC process enough so that it behaves like a singular virtual machine from the perspective of both the model owner and the end user. Once this abstraction is developed and standardized, model serving will be similar to that of traditional MLOps pipelines, since the underlying privacy-preserving operations are abstracted and no longer interfere with the deployment process.

Model monitoring is a process that provides data about the quality of the model being used in production. This data can then be potentially used as a trigger to retrain/redevelop the model, as models are inherently decaying over time due to the availability of new data or shifting of the trends in data.If the model inference is to be run with SMPC, then this step is impossible to complete, as the result of the inference should only be visible to the end user. A solution to this could be monitoring of a copy of the model conducted within the domain of the model owner.

The final part that needs to be addressed regarding the MLOps pipeline as well as machine learning model design is the actual model architecture and the experimental phase of model development. In the previous chapter, it was shown that even with a much simpler model structure similar or even better qualitative results can be achieved, while having the lowest inference times. The simpler architecture has even more impact on ppML, as the inference time in the encrypted domain has become a critical metric to keep track of. Machine learning experts should strive for simplicity and very specifically designed models for the problem they address if optimal performance in the encrypted domain is to be achieved.

7. Conclusions

Our findings demonstrate that a purely data-driven learning approach achieves competitive accuracy in predicting energy consumption, particularly when using manually engineered features such as joint angles, velocities, and accelerations. Among the evaluated model architectures, dense networks that used all available features delivered the best results in terms of qualitative metrics and inference speed, both in plaintext and encrypted settings.

We further integrated privacy-preserving machine learning (ppML) techniques based on secure multi-party computation (SMPC) to enable secure inference on proprietary data without revealing either the input or the trained model weights. Our analysis revealed that model architecture significantly affects inference performance in the encrypted domain, with dense networks being more efficient than convolutional or recurrent architectures. We also evaluated optimization strategies such as plaintext weight permutation and plaintext activation evaluations, showing that plaintext weight permutation can significantly reduce inference times but with trade-offs in privacy.

Furthermore, we examined the implications of ppML integration within modern machine learning operations (MLOps) workflows. Most MLOps stages remain unaffected, but our analysis identified two areas, namely model serving and model monitoring, that require some adaptations. More specifically, serving privacy-preserving models demands careful coordination between model owners, end-users, and secure computation services, while traditional monitoring must be redesigned since model outputs are not directly observable under SMPC.

This work provides a framework for secure, data-driven energy prediction in industrial robotics. It contributes insights for both the design of efficient ppML models and the adaptation of MLOps pipelines, supporting adoption of ppML in increasingly complex industrial ecosystems.

Future work might focus on developing and evaluating more complex modeling tasks in industrial settings (different robot configurations, effect of a loaded arm on the model, etc.), potentially giving more insights on ppML for a different set of machine learning architectures. It would be also interesting to streamline the performance evaluation process, be it by researching novel means to support software developers, by estimating expected performance for given models, or by providing automated guidance for optimization. Another focus might be deploying and testing our framework in a live setting and evaluating it with realistic network complications. Investigating the potential integration of industrial standards for communication (e.g., OPC UA) into the ppML pipeline is another potential extension in the future.

Author Contributions: Conceptualization, P.S., A.S., T.L., S.H. and R.H.; methodology, P.S., A.S., T.L., S.H. and R.H.; software, A.S. and P.S.; validation, A.S. and P.S.; formal analysis, P.S.; investigation, A.S. and P.S.; data curation, P.S.; writing—original draft preparation, A.S. and P.S.; writing—review and editing, A.S., P.S., T.L., S.H. and R.H.; visualization, A.S. and P.S.; supervision, T.L., S.H. and R.H.; All authors have read and agreed to the published version of the manuscript.

Funding: This work has received funding from the EU Horizon Europe Work Programme under project QSNP (no. 101114043) and the Austrian Research Promotion Agency FFG within the PRESENT project (grant no. 899544).

Institutional Review Board Statement: Not applicable.

Data Availability Statement: Dataset available on request from the authors.

Conflicts of Interest: The authors declare no conflicts of interest.

References

1. Abdoune, F.; Ragazzini, L.; Nouiri, M.; Negri, E.; Cardin, O. Toward Digital Twin for Sustainable Manufacturing: A Data-Driven Approach for Energy Consumption Behavior Model Generation. *Comput. Ind.* **2023**, *150*, 103949. [CrossRef]
2. Sekala, A.; Blaszczyk, T.; Foit, K.; Kost, G. Selected Issues, Methods, and Trends in the Energy Consumption of Industrial Robots. *Energies* **2024**, *17*, 641. [CrossRef]
3. Steurer, P.; Skuta, A.; Hegenbart, S.; Hoch, R.; Loruenser, T. Data-Driven Modeling for Privacy-Preserving Energy Prediction of Industrial Robots. In Proceedings of the 5th Winter IFSA Conference on Automation, Robotics & Communications for Industry 4.0/5.0, ARCI' 2025, Granada, Spain , 19–21 February 2025; Yurish, S., Ed.; pp. 184–189. [CrossRef]
4. Walther, J.; Weigold, M. A Systematic Review on Predicting and Forecasting the Electrical Energy Consumption in the Manufacturing Industry. *Energies* **2021**, *14*, 968. [CrossRef]
5. Liu, A.; Liu, H.; Yao, B.; Xu, W.; Yang, M. Energy consumption modeling of industrial robot based on simulated power data and parameter identification. *Adv. Mech. Eng.* **2018**, *10*, 1687814018773852 . [CrossRef]
6. Yan, K.; Xu, W.; Yao, B.; Zhou, Z.; Pham, D.T. Digital Twin-Based Energy Modeling of Industrial Robots. In *Methods and Applications for Modeling and Simulation of Complex Systems, Proceedings of the 18th Asia Simulation Conference, AsiaSim 2018, Kyoto, Japan, 27–29 October 2018*; Communications in Computer and Information Science; Li, L., Hasegawa, K., Tanaka, S., Eds.; Springer: Singapore, 2018; pp. 333–348. [CrossRef]

7. Kajzr, D.; Myslivec, T.; Černohorský, J. Modelling, Analysis and Comparison of Robot Energy Consumption for Three-Dimensional Concrete Printing Technology. *Robotics* **2024**, *13*, 78. [CrossRef]
8. Swevers, J.; Verdonck, W.; De Schutter, J. Dynamic Model Identification for Industrial Robots. *IEEE Control Syst. Mag.* **2007**, *27*, 58–71 . [CrossRef]
9. Wu, J.; Wang, J.; You, Z. An overview of dynamic parameter identification of robots. *Robot. Comput.-Integr. Manuf.* **2010**, *26*, 414–419. [CrossRef]
10. Bargsten, V.; Zometa, P.; Findeisen, R. Modeling, parameter identification and model-based control of a lightweight robotic manipulator. In Proceedings of the 2013 IEEE International Conference on Control Applications (CCA), Hyderabad, India, 28–30 August 2013; pp. 134–139. ISSN 1085-1992. [CrossRef]
11. Paryanto; Brossog, M.; Bornschlegl, M.; Franke, J. Reducing the energy consumption of industrial robots in manufacturing systems. *Int. J. Adv. Manuf. Technol.* **2015**, *78*, 1315–1328. [CrossRef]
12. Zhang, M.; Yan, J. A data-driven method for optimizing the energy consumption of industrial robots. *J. Clean. Prod.* **2021**, *285*, 124862. [CrossRef]
13. Yin, S.; Ji, W.; Wang, L. A machine learning based energy efficient trajectory planning approach for industrial robots. *Procedia CIRP* **2019**, *81*, 429–434. [CrossRef]
14. Lorünser, T.; Wohner, F.; Krenn, S. A Privacy-Preserving Auction Platform with Public Verifiability for Smart Manufacturing. In Proceedings of the ICISSP 2022, Online, 9–11 February 2022; Mori, P., Lenzini, G., Furnell, S., Eds.; SCITEPRESS: Setúbal, Portugal , 2022; pp. 637–647. [CrossRef]
15. Lorünser, T.; Wohner, F.; Krenn, S. A Verifiable Multiparty Computation Solver for the Linear Assignment Problem: And Applications to Air Traffic Management. In Proceedings of the CCSW 2022, Los Angeles, CA, USA, 7–11 November 2022; Regazzoni, F., van Dijk, M., Eds.; ACM: New York, NY, USA, 2022; pp. 41–51. [CrossRef]
16. Keller, M. MP-SPDZ: A Versatile Framework for Multi-Party Computation. In Proceedings of the 2020 ACM SIGSAC Conference on Computer and Communications Security, Virtual, 9–13 November 2020.
17. Ma, J.; Zheng, Y.; Feng, J.; Zhao, D.; Wu, H.; Fang, W.; Tan, J.; Yu, C.; Zhang, B.; Wang, L. SecretFlow-SPU: A Performant and User-Friendly Framework for Privacy-Preserving Machine Learning. In Proceedings of the USENIX ATC '23, Boston, MA, USA, 10–12 July 2023; pp. 17–33.
18. Mohassel, P.; Rindal, P. ABY3: A Mixed Protocol Framework for Machine Learning. In Proceedings of the 2018 ACM SIGSAC CCS, CCS '18, Toronto, ON, Canada, 15–19 October 2018; pp. 35–52. [CrossRef]
19. Cramer, R.; Damgård, I.; Escudero, D.; Scholl, P.; Xing, C.; $SPD\mathbb{Z}_{2^k}$: Efficient MPC mod 2^k for Dishonest Majority. In Advances in Cryptology – CRYPTO 2018, Proceedings of the 38th Annual International Cryptology Conference, Santa Barbara, CA, USA, August 19—23, 2018; Springer: Berlin/Heidelberg, Germany, 2018; Volume II, pp. 769–798. [CrossRef]
20. Huang, Z.; Lu, W.j.; Hong, C.; Ding, J. Cheetah: Lean and Fast Secure Two-Party Deep Neural Network Inference. In Proceedings of the 31st USENIX Security Symposium (USENIX Security 22), Boston, MA, USA, 10–12 August 2022 .
21. Knott, B.; Venkataraman, S.; Hannun, A.; Sengupta, S.; Ibrahim, M.; van der Maaten, L. CrypTen: Secure Multi-Party Computation Meets Machine Learning. In Proceedings of the Advances in Neural Information Processing Systems 34 (NeurIPS 2021), Online, 6–14 December 2021.
22. Pub MA KUKA Sunrise Cabinet (PDF) en. Available online: https://www.oir.caltech.edu/twiki_oir/pub/Palomar/ZTF/KUKARoboticArmMaterial/MA_KUKA_Sunrise_Cabinet_en.pdf (accessed on 27 August 2025).
23. Lynch, K.M.; Park, F.C. *Modern Robotics: Mechanics, Planning, and Control*; Cambridge University Press: Cambridge, UK, 2017.
24. Birchfield, S. *Image Processing and Analysis*, 1st ed.; Cengage Learning: Mason, OH, USA, 2016.
25. Kingma, D.P.; Ba, J. Adam: A Method for Stochastic Optimization. In Proceedings of the 3rd International Conference on Learning Representations, ICLR 2015, San Diego, CA, USA, 7–9 May 2015; Conference Track Proceedings; Bengio, Y., LeCun, Y., Eds.; 2015.
26. Krizhevsky, A.; Sutskever, I.; Hinton, G.E. ImageNet Classification with Deep Convolutional Neural Networks. In Proceedings of the Advances in Neural Information Processing Systems 25 (NIPS 2012), Lake Tahoe, NV, USA, 3–6 December 2012; Pereira, F., Burges, C., Bottou, L., Weinberger, K., Eds.; Curran Associates, Inc.: Red Hook, NY, USA, 2012; Volume 25.
27. Lecun, Y.; Bottou, L.; Bengio, Y.; Haffner, P. Gradient-based learning applied to document recognition. *Proc. IEEE* **1998**, *86*, 2278–2324. [CrossRef]
28. Luo, J.; Chen, G.; Zhang, Y.; Liu, S.; Wang, H.; Yu, Y.; Zhou, X.; Qi, Y.; Xu, Z. Centaur: Bridging the Impossible Trinity of Privacy, Efficiency, and Performance in Privacy-Preserving Transformer Inference. *arXiv* **2024**, arXiv:2412.10652. [CrossRef]
29. Luo, J.; Zhang, Y.; Zhang, Z.; Zhang, J.; Mu, X.; Wang, H.; Yu, Y.; Xu, Z. SecFormer: Fast and Accurate Privacy-Preserving Inference for Transformer Models via SMPC. *arXiv* **2024**, arXiv:2401.00793. [CrossRef]
30. Li, D.; Shao, R.; Wang, H.; Guo, H.; Xing, E.P.; Zhang, H. MPCFormer: Fast, performant and private Transformer inference with MPC. *arXiv* **2023**, arXiv:2211.01452. [CrossRef]

31. Lorünser, T.; Wohner, F. Performance Comparison of Two Generic MPC-frameworks with Symmetric Ciphers. In Proceedings of the ICETE (2), Paris, France, 8–10 July 2020; pp. 587–594.
32. Google MLOps Archtecture. Available online: https://cloud.google.com/architecture/mlops-continuous-delivery-and-automation-pipelines-in-machine-learning (accessed on 1 July 2025).

machines

MDPI

Article

An Echo State Network-Based Light Framework for Online Anomaly Detection: An Approach to Using AI at the Edge

Andrea Bonci [1], Renat Kermenov [1], Lorenzo Longarini [1], Sauro Longhi [1], Geremia Pompei [2], Mariorosario Prist [1,*] and Carlo Verdini [3]

[1] Department of Information Engineering (DII), Polytechnic University of Marche, 60131 Ancona, Italy; a.bonci@univpm.it (A.B.); r.kermenov@pm.univpm.it (R.K.); s1110740@studenti.univpm.it (L.L.); s.longhi@univpm.it (S.L.)

[2] Institute of Chemical and Physical Processes of National Research Council, 56124 Pisa, Italy; geremia.pompei@pi.ipcf.cnr.it

[3] Syncode Scarl, 93900 Fermo, Italy; carlo.verdini@syncode.it

* Correspondence: m.prist@staff.univpm.it

Abstract: Production efficiency is used to determine the best conditions for manufacturing goods at the lowest possible unit cost. When achieved, production efficiency leads to increased revenues for the manufacturer, enhanced employee safety, and a satisfied customer base. Production efficiency not only measures the amount of resources that are needed for production but also considers the productivity levels and the state of the production lines. In this context, online anomaly detection (AD) is an important tool for maintaining the reliability of the production ecosystem. With advancements in artificial intelligence and the growing significance of identifying and mitigating anomalies across different fields, approaches based on artificial neural networks facilitate the recognition of intricate types of anomalies by taking into account both temporal and contextual attributes. In this paper, a lightweight framework based on the Echo State Network (ESN) model running at the edge is introduced for online AD. Compared to other AD methods, such as Long Short-Term Memory (LSTM), it achieves superior precision, accuracy, and recall metrics while reducing training time, CO_2 emissions, and the need for high computational resources. The preliminary evaluation of the proposed solution was conducted using a low-resource computing device at the edge of the real production machine through an Industrial Internet of Things (IIoT) smart meter module. The machine used to test the proposed solution was provided by the Italian company SIFIM Srl, which manufactures filter mats for industrial kitchens. Experimental results demonstrate the feasibility of developing an AD method that achieves high accuracy, with the ESN-based framework reaching 85% compared to 80.88% for the LSTM-based model. Furthermore, this method requires minimal hardware resources, with a training time of 9.5 s compared to 2.100 s for the other model.

Keywords: Industry 4.0; anomaly detection; machine learning; predictive maintenance; echo state network; edge computing

1. Introduction

Nowadays, industries are increasingly attracted to Machine Learning (ML), as it is a powerful tool that allows for the conversion of large amounts of data into potential predictions or the extraction of other useful information. This is reflected in a significant contribution to all decision-making processes based on data [1]. For a long time, manufacturing companies have aimed to improve quality and efficiency by adopting new methodologies such as energy optimization [2–4], diagnostics [5], modeling of complex behaviors such as productive interactions [6–8], human and complex dynamics interactions [9], and even the integration of systems based on cyber-physical architectures [10] and interconnection of systems [11,12]. Recently, there has been growing interest in ML, justified by its versatility, as it allows for the reduction or elimination of redundant activities, the resolution of existing problems, and the extraction of hidden value through data analysis and pattern recognition. Applications of ML in industrial facilities for tasks such as classification, continuous estimation, clustering, process optimization, and generation of data allow us to address a lot of industrial problems [13].

A particularly relevant aspect is anomaly detection (AD), which focuses on identifying anomalous or unusual data patterns. This approach allows problems to be anticipated and solved before they can negatively impact operations, thus helping to improve the stability and efficiency of production systems. Some examples are as follows. Tanuska et al. [14] introduced a smart approach using an Industrial Internet of Things (IIoT) network based on Long-Range Wide Area Network (LoRaWAN) technology to solve the problem of accidental carrier bearings damage to an assembly conveyor, while Tancredi et al. [15] developed a Digital Twin (DT) architecture as a replica of a physical system (a tube-in-tube indirect machine for fluid food pasteurization) to support the technique of predicting system parameters and verifying the effectiveness of the production process.

Currently, the main ML methods for AD rely on generative models such as Generative Adversarial Networks (GANs), a class of artificial intelligence algorithms used in unsupervised machine learning. GANs are implemented by a system of two neural networks contesting each other in a zero-sum game, where one network, the generator, tries to create data indistinguishable from real data, while the other network, the discriminator, aims to accurately distinguish between real and generated data [16,17]. Other models include Variational AutoEncoders (VAEs), an advanced class of neural network models that integrate principles of variational inference with deep learning to perform supervised and unsupervised learning tasks [18,19], and Graph AutoEncoders (GAEs), a specialized form of autoencoders designed to work with graph-structured data. GAEs encode the nodes of a graph into a latent, low-dimensional space while preserving the graph's topological structure, then reconstruct the graph or predict its properties from these embeddings [20,21]. Finally, time-series forecasting can also be mentioned, which uses a recurrent neural network (RNN), a neural network with loops that allows information to persist from one step to the next, or Long Short-Term Memory (LSTM), a specialized RNN developed to mitigate the vanishing and exploding gradient problems [22–24] and widely used to predict future values in the context of AD in smart manufacturing [25,26]. Within the class of RNNs, another type of model is the Echo State Network (ESN), which has been used in industrial contexts. For example, Wang [27] combined the ESN model and sparse ADABOOST methodology to forecast industrial electricity consumption in China, while Pang presented a neurodynamic approach to Model Predictive Control (MPC) of unknown nonlinear dynamical systems using two ESN-based models [28]. Conversely, Martino [29] focused on forecasting the blast furnace gas amount, its heating power, and the blast furnace gas demand by hot blast stoves for integrated steelworks.

AD needs to be evaluated differently depending on the model used; a single metric cannot be applied to all models. In the context of generative models, AD is calculated as the difference between the observed data and the outcomes predicted by the model, using the loss function as an index to measure the mismatch. However, in the context of forecasting methods, it is more meaningful to measure this difference using indicators such as the Mean Squared Error (MSE) [30] or the Mean Absolute Error (MAE) [31]. Currently, most studies on AD prioritize model performance over computational limitations when applying these models to edge devices. Furthermore, the datasets used in these studies usually come from standard online sources, which are often based on vibration data [32–34], rather than from devices that operate with real-time feedback. Furthermore, considering that acquisition devices create massive amounts of data, it is crucial to implement appropriate algorithms and even artificial intelligence (AI) techniques as locally as possible. In fact, the use of only cloud processing could make data management difficult, leading to bandwidth congestion, privacy issues, high latency, and lack of reliability [35]. In response to these limitations, an alternative approach using the ESN, optimized to run on resource-constrained platforms at the edge of the network, has been proposed. The LSTM model was used for comparison because it is one of the most widely used models in industrial AD benchmarks [36]. At the same time, there are many other valid state-of-the-art models, such as GRUs and Transformer-based architectures. In fact, as described by Braei et al. [37], the most popular approaches used for anomaly detection based on univariate and multivariate time series include Multi-Layer Perceptron (MLP), CNNs, RNNs, LSTM networks, Gated Recurrent Units (GRUs), and autoencoders. By comparing the performance of these models using a dataset composed of univariate time series that contain contextual anomalies, the authors demonstrate that LSTM outperforms the other methods not only in terms of accuracy but also in terms of inference time. Since the aim of the present work is to evaluate the use of the ESN as a viable solution for AD, and considering that, based on the results highlighted in [37], LSTM networks currently yield the best performance for such tasks, this study uses LSTM networks as a reference model for evaluating the proposed solution. In particular, an experimental setup on a Raspberry Pi 4 using historical electricity consumption data from a smart meter connected to a metal filter production machine has been built. The results show that the proposed AD method provides 4% higher accuracy than LSTM networks while dramatically reducing the processing time from 35 min to just 9.5 s. In conclusion, the contributions of the proposed solution can be summarized as follows:

- Lightweight ESN-based anomaly detector at the edge: Executing AD at the edge ensures real-time analysis and an immediate response to potential disruptions, which are crucial in time-sensitive production environments. This near-instantaneous feedback loop allows for prompt corrective measures, minimizing the likelihood of extended downtime and ensuring continuous production flow.
- Addressing the edge computing paradigm: The proposed solution allows for the use of low-specification embedded device hardware like a Raspberry Pi, equipped with only 1 Gb RAM, increasing the economic benefits while reducing power consumption during training and inference. In fact, instead of preparing and compressing all data for transmission to the cloud, edge computing allows for the processing of data for training and inference on a local device located near the data sources. This proximity reduces the time to transmit data over a network, reduces the computational resources required, and lowers costs.
- Continual learning: Considering that the proposed AD architecture is based on the ESN model and runs on an embedded device, it is possible to continuously update the model at the edge to maintain its relevance and performance despite evolving operational conditions. This approach can be viewed as a continual learning technique, similar to a replay-based method [38], where periodically revisiting and integrating past and new data during training enhances the preservation of previous knowledge and strengthens the model's robustness.
- Industrial application: Comprehensive testing is carried out within a practical environment, an automatic press machine used to produce metal filters provided by the Italian company SIFIM, to assess the efficacy of the proposed method. The dataset comprises time-series data from eighty-eight variables related to energy parameters, selected from the three hundred available through commercially obtainable industrial-grade smart metering IIoT sensors, to replicate our investigative procedure. The outcomes indicate that the proposed approach offers enhanced promptness in error identification and achieves superior predictive precision.
- Performance: The experimental results using an edge device on a real production machine show that, compared to traditional time-series models (LSTM), the ESN-based framework proposed in this paper achieves very good performance.

This paper is organized as follows. Section 2 introduces the materials and methods for the proposed AD framework, highlighting the model based on the ESN architecture, the anomaly detector, and the system architecture, including the pilot case, the smart meter module used to gather data from the pilot test, and the data preparation. The experimental analysis based on a real production machine, along with a visualization of the results and a comparison using different models, is discussed in Section 3. Discussions and boundary conditions are presented in Section 4. Finally, Section 5 presents the conclusions and future improvements.

2. Material and Methods

This section provides a general overview of the methodology adopted to address AD using ESN, the hardware and software architecture proposed to implement the solutions, and the methods adopted to evaluate the considered models.

2.1. Proposed Solution

The proposed framework is based on the ESN model, and the results are compared with those of the LSTM-based approach. This section presents an overview of the ESN-based model and LSTM-based model, along with a clear description of the proposed architecture.

2.1.1. Echo State Network

The ESN-based model [39] is a special type of implementation of an RNN based on the reservoir computing paradigm [40,41]. The model can be divided into two main components: the first is the randomly initialized hidden layer called a reservoir, which is not trained, and represents the recurrent part that captures the temporal dependencies of time series, and the second is a trainable readout, which is responsible for mapping the high-dimensional dynamic states of the reservoir to the desired output (see Figure 1).

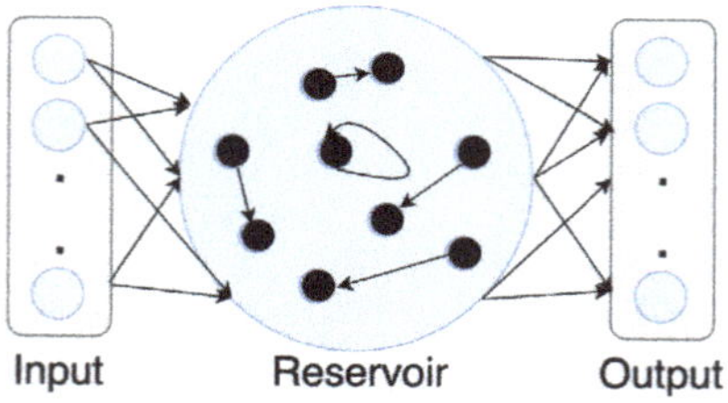

Figure 1. ESN diagram.

2.1.2. Long-Short Term Memory

The LSTM-based model [42] is a particular type of implementation of a recurrent neural network that uses predefined cells to learn both short-term and long-term temporal dependencies. These cells, called gates, enable the LSTM to avoid the issues of gradient vanishing or gradient exploding that affect standard recurrent neural networks. Figure 2 illustrates the fundamental blocks of LSTM, where information is managed by three gates.

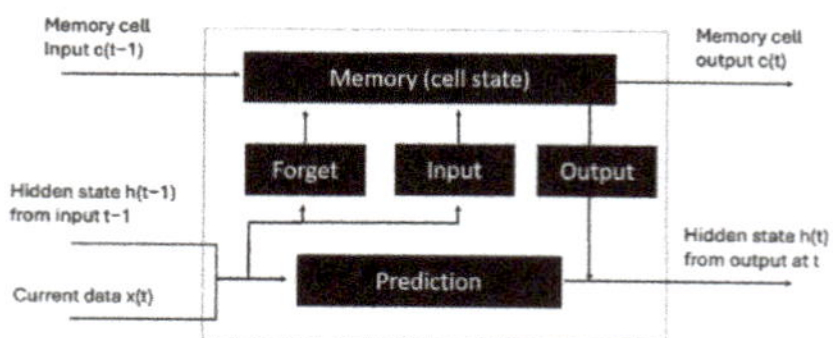

Figure 2. LSTM diagram.

The first gate, the input gate, controls the value to be added and stored. The second gate, the forget gate, is responsible for maintaining or discarding specific information, while the third gate, the output gate, is responsible for updating the cell state by combining the previous hidden state with new input data.

2.1.3. AD Framework

The AD framework can be divided into two different parts:

1. Next-step prediction model: This part is related to a deep learning model that performs the next-step prediction task by emulating the behavior of the time series.
2. Anomaly detector: This part is related to the error-based method for understanding anomalies by comparing the statistics of the error in the training set with those computed during the inference phase.

From an operational point of view, the following three main phases can be taken into account:

1. Training phase: This is the learning phase of the model. The model, initialized randomly, is adapted to solve the task of predicting the next step by updating the internal weights to perform the current task.
2. Training error SD σ-computation: This involves the application of the MSE between the prediction and target values to retrieve the standard deviation σ of the error in the training set. As highlighted in Figure 3, the sequence $x(t-1), x(t), x(t+1), \ldots$ etc., represents the sequential input of data over time, which is the time-series data input of the model. The model function processes the input data and generates predictions of the next data points in the sequence or the prediction of the next step. The MSE is calculated between the next step's prediction and the target values in order to produce error values to be collected. The goal of this step is to produce the standard deviation σ of the errors in the training set.
3. Inference phase: In the inference phase, the trained model is able to sequentially take the input data for each time step $x(t)$ and provide the next-step prediction values $x'(t+1)$. The anomaly detector computes the error between the prediction $x'(t+1)$ and the target values $x(t+1)$ using the MSE function (see Figure 4).

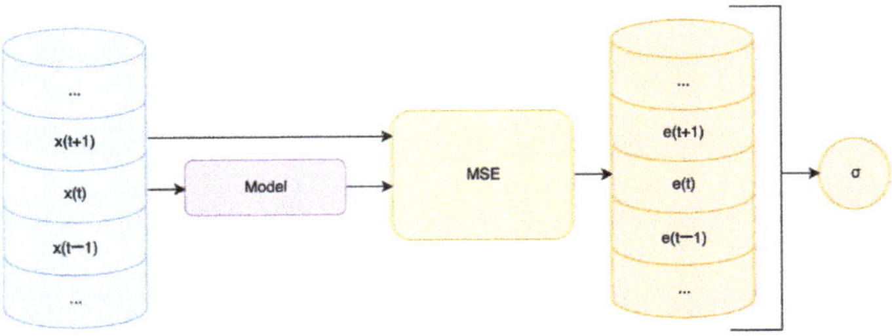

Figure 3. Architecture for the computation of the standard deviation of the error in the training set.

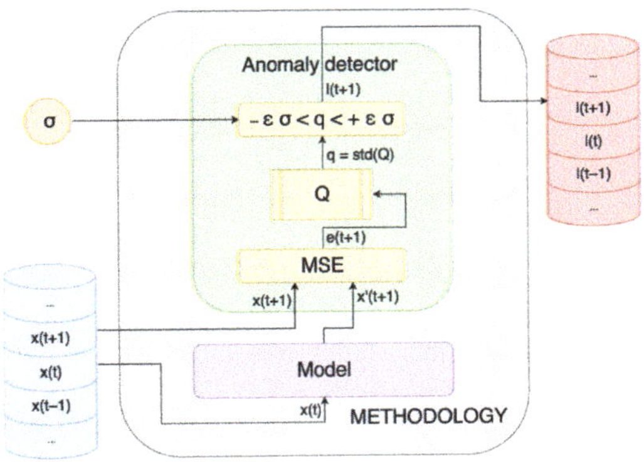

Figure 4. Framework architecture for the inference phase.

A queue Q is able to collect $|Q|$ recent errors. The size of the queue is chosen based on a model selection phase that tests the metric results on a validation set. Then, the standard deviation of the stored errors is computed as $q = std(Q)$, providing a vector of values that can be compared with those computed previously in the training set. The condition that allows for normal behavior is $-\epsilon\sigma < q < +\epsilon\sigma$, where ϵ is chosen during model selection $\epsilon|\epsilon > 0, \epsilon \in 0.8, 1, 1.2$. If the previous condition is true, the output value is 0 to indicate that there are no anomalies, while in the opposite case, the output is 1 to indicate an anomaly. The results are stored in a buffer containing labels (0 or 1) for each time step. The choice of values for ϵ (0.8, 1.0, and 1.2) is based on empirical findings. These values represent a balanced compromise for anomaly detection, minimizing excessive adaptation to noise. Specifically, the selected range allows for the identification of areas where errors may manifest prior to the model detecting an anomaly. The values correspond to ratios of the standard deviation of the error during training, with empirical experimentation supporting their effectiveness in this context.

The goal of this work is to obtain a neural network with state-of-the-art efficiency while maintaining a low computational burden. In this case, a valid competitor for the LSTM model is the ESN model. The ESN is a recurrent model that leverages dynamical systems theory to map an input time series to a trajectory in higher dimensionality without diverging to simplify the classification or regression task. The ESN is part of the family of reservoir computing, which, as mentioned above, is a framework that consists of recurrent models split into two parts: a reservoir and a readout. The reservoir is the recurrent part that is untrained, while the readout is the only trained part that maps the projection of the reservoir to the output prediction. The ESN is very efficient because it does not use the backpropagation through time (BPTT) training algorithm to train it but rather a one-shot learning function called ridge regression. The model selection phase is important for choosing the best configuration of hyperparameters. In particular, we focus on choosing the optimal queue length $|Q|$ that, unlike state-of-the-art works (see e.g., [43]), is not fixed to a constant value. In this way, we select $|Q|$ with respect to the best performance achieved on a validation set.

2.2. System Architecture

The system architecture for the proposed use case is illustrated in Figure 5. It is composed of four main parts:

- Edge Device: A low-computational-resources hardware device.
- Pilot case: An automated production machine dedicated to the manufacture of metal filters.
- Smart Meter: An Industrial Internet of Things device equipped with sophisticated three-phase system network analyzers.
- User Interface: A visualization module with a high-level processing system.

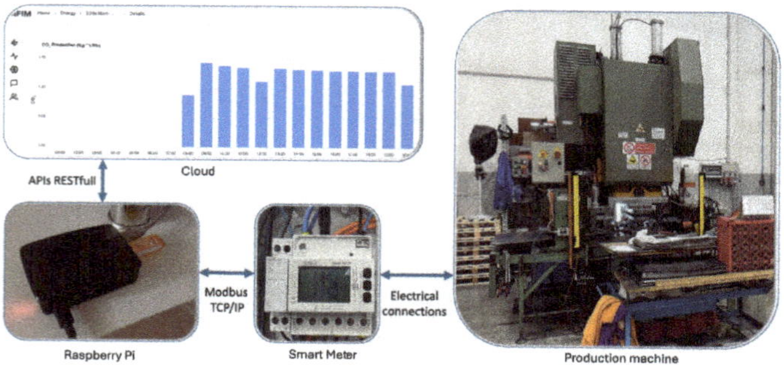

Figure 5. Architecture diagram.

2.2.1. Device at the Edge

The validation phase of the results was performed through experimentation on a real device installed at the edge, with data coming from a real environment. The device considered was a Raspberry Pi 4 model B as the hardware platform, equipped with a 64-bit quad-core CPU running at 1.5 GHz and 1 GB of RAM. Although not specifically designed as an edge-only system, this device was chosen for its connectivity, flexibility, and ability to quickly implement complex control algorithms (such as [44]), especially in scenarios requiring near-real-time processing. The framework was developed entirely in Python, which allows for effective prototyping even for advanced applications, provided it is supported by a suitable operating system [45].

2.2.2. Pilot case

Thanks to the infrastructure made available by the Italian company SIFIM Srl, it was possible to set up a pilot case that allowed for testing the proposed approach in a real context and evaluating its performance. In particular, the infrastructure consisted of an automated production machine dedicated to the manufacture of metal filters (see Figure 6), where products can be localized with novel and fast-tracing systems such as those in [46].

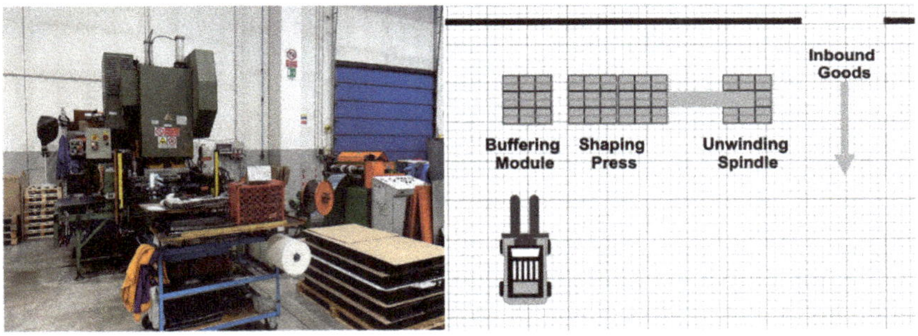

Figure 6. Production machine and production layout.

The production machine is used to transform steel or aluminum strips into shaped metal mesh filters used in commercial kitchen hoods. The machine is composed of three major components: an unwinding spindle with a coil, a shaping press, and a buffering module. These components are described as follows:

- Unwinding Spindle: This component is the starting point of the filter production process, where the raw material (steel or aluminum strip) is loaded onto the machine in relation to the article to be produced.
- Shaping Press: This is the main component of the machine, composed of a hydraulic shaping press. It receives the metal strip from the unwinding spindle and applies specific force and shape to the strip in relation to the article to be produced, like a non-nesting system of smooth-walled honeycombs.
- Buffering Module: This module is responsible for temporarily holding the cut filter mats on a flat surface, allowing for smooth transitions between the cutting process and subsequent handling.

The first section, which is designed solely for unwinding the material, is equipped with a 3 kW three-phase asynchronous motor, while the main module, the hydraulic press, utilizes an 11.5 kW three-phase asynchronous motor. The machine also includes smaller cutting motors to manage only the horizontal and vertical movements of the press. The operation of a press machine for metallic filter fabrication involves the application of significant force to shape steel or aluminum strips into filters. This process inherently subjects various components of the machine, particularly the shaping press, to mechanical stress. The repeated application of force, along with the continuous motion of the unwinding and winding spindles, results in the wear and tear of mechanical parts over time. Components such as gears, bearings, and the press surface are especially susceptible to stress-related degradation, which can affect the machine's precision and overall performance. Given the mechanical stress involved in the operation of a metallic filter press machine, predictive maintenance using the AD approach becomes very important to ensure longevity, reliability, and efficiency.

2.2.3. Smart Meter Module

The S604 external module from Seneca [47] allows for the electrical parameters of the production machine to be monitored. This instrument, equipped with sophisticated three-phase network analyzers, makes it possible to accurately collect and store the energy data needed to analyze and optimize the production process. Indeed, the S604 can replace single analog meters as well as multifunction meters like voltmeters, ammeters, wattmeters, varmeters, frequency meters, phase meters, and energy meters. The smart meter can record a multivariate vector with 88 different measurements at a sample rate of 1 Hz, including the following:

- Line voltage V1, V2, and V3; phase voltage V12, V13, and V23; and phase current A1, A2, and A3.
- Phase active power P1, P2, and P3; phase apparent power S1, S2, and S3; and phase reactive power Q1, Q2, and Q3.
- Phase power factor PF1, PF2, and PF3; and phase displacement power factor DPF1, DPF2, and DPF3.
- Line total harmonic distortion of the voltage waveform THDV1, THDV2, and THDV3; phase total harmonic distortion of the voltage waveform THDV12, THDV13, and THDV23; and total harmonic distortion of the current waveform THDA1, THDA2, THDA3, and THDAN.
- Harmonic components of phase current HaA1 from 0th to 15th; HaA2 from 0th to 15th; and HaA3 from 0th to 15th.
- Frequency.

2.2.4. User Interface

The S604 module is integrated into the system by connecting directly to the production machine's three-phase connections. Communication with the Raspberry Pi is ensured by the Modbus TCP/IP protocol, which allows for efficient data transmission. The results of the AD analysis are then transferred to the cloud, where they undergo further processing. Alerts and reports can be generated from this process, such as carbon dioxide production levels in kilograms per kilowatt-hour (Kg/kWh) (see Figure 7) across various times of the day [48].

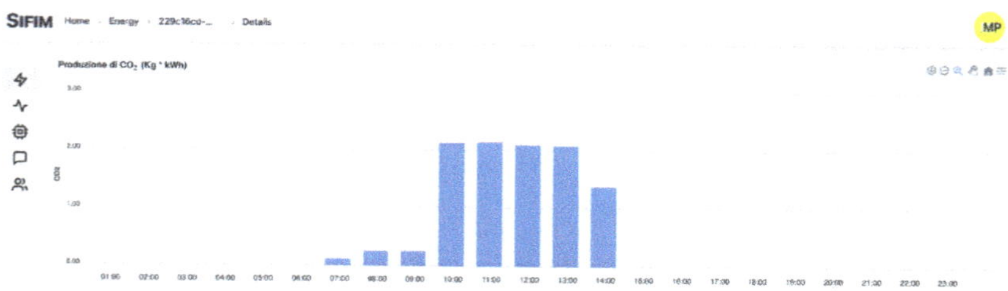

Figure 7. Cloud dashboard—CO_2 production.

2.3. *Model Evaluation*

This subsection presents how the model was evaluated by describing how the data was prepared, the metrics used to select the model, and the selection phase.

2.3.1. Data preparation

The embedded device implementing the proposed architecture was placed in the production environment. Five different approaches were applied to prepare the data [49] in the correct way, i.e., transforming it into a format suitable for the analysis to be performed. In this way, we ensured that the model received accurate data from the field. The approaches applied were the following:

- Data rectification: When the industrial machine is in a non-operational mode, or turned off, the system, which is designed to gather information about the machine state, records null as the data value. This null value indicates the absence of data, which is an effective measure to reflect that no production activity is occurring during this period and, consequently, no relevant data on parameters such as energy consumption, output, and performance metrics can be gathered.
- Data cleaning: Data cleaning is an essential procedure in data management that involves refining datasets to ensure that they are accurate, consistent, and ready for analysis. In this specific context, there is useless information such as serial numbers, identification numbers (IDs), and timestamps, which, in this analytical scenario, do not contribute meaningful insights. Due to the fact that the analysis aims to implement an AD methodology, these useless data points are unused and hence removed to streamline the dataset. This refinement allows for a more focused analysis of variables that directly impact operational efficiency.
- Data normalization: By applying min-max normalization, all features are translated into a uniform scale ranging from 0 to 1. Here, x represents the original value, min is the minimum value in the dataset, and max is the maximum value. This transformation maintains the relationships in the data while standardizing the range. Normalization ensures that each input feature contributes equally to the analysis, allowing the model to process data without any implicit bias that might arise from the scale of the variables. In addition, large values in the data can cause numerical problems during the computation process. Normalization mitigates the risk of encountering numerical instabilities by scaling values to a limited and standardized range.
- Dataset splitting: The dataset D consists of 269 time series with 300 timesteps and 88 features. For the purpose of constructing an effective and robust model, the dataset D is partitioned into three distinct subsets: the training set D_{train} (60%), the validation set D_{valid} (20%), and the test set D_{test} (20%). This partitioning is guided by the principle of representativeness and the sequential nature of time-series data, ensuring that the temporal integrity of the dataset is maintained.
- Data Labeling: In the validation and test sets, the acquired occurrences of faults are kept. Class 0 is assigned to the non-fault cases and class 1 is assigned to the fault cases.

2.3.2. Metrics

The metrics used to quantify the results can be classified into two categories: metrics that evaluate effectiveness and metrics that assess efficiency. They are described as follows:

- Effectiveness metrics: All effectiveness metrics are applied to the test set to evaluate the overall methodology in performing a binary classification task. Consequently, the primary metrics used in binary classification scenarios include accuracy, precision, recall, and F1 score.
- Efficiency metrics: The metrics employed to assess the efficiency of the methodology encompass time-related measures and CO_2 emissions. The time metric evaluates the speed of the model during both the training and inference phases. Additionally, CO_2 emissions are quantified during these phases, indicating the volume of carbon dioxide produced (in kg/CO_2) throughout the training and inference processes. Specifically, these emissions are measured using the CodeCarbon Python library.

2.3.3. Model Selection

The hyperparameters for the ESN model were determined using the trial-and-error method. This approach effectively yields satisfactory results by exploring various combinations of hyperparameters until an acceptable level of error reduction is achieved, which is a common practice in hyperparameter optimization. This phase, referred to as model selection, involves initializing and training new models on the training set with the selected hyperparameter combinations and subsequently evaluating their performance on the validation set. In the context of binary classification, rather than employing an error minimization technique, the accuracy metric is utilized to select the optimal hyperparameter configuration that maximizes classification performance (see Table 1).

Table 1. ESN model selection.

Model	Hyperparameter	Value	Description
ESN	reservoir_size	[100, 200 *, 300]	Reservoir and hidden state size
	alpha	[0.2, 0.5, 0.8 *]	Leaky rate hyperparameter
	input_ratio	[0.7, 0.9 *]	Hyperparam. to scale input weights
	model_spectral_radius	[1.2, 0.9 *]	Spectral radius hyperparam.
	model_input_sparsity	[0.5 *]	Input weights' sparsity
	reservoir_sparsity	[0.9 *]	Reservoir weights' sparsity
	model_regularization	[0.001, 0.01 *, 0.0001]	Thikonov regularization hyperparam.
	n_layers	[1 *]	Number of reservoir layers
	washout	[100 *]	To cut first hidden states
	ϵ	[0.8 *, 1, 1.2]	ϵ constant of anomaly detector (*ad*)
	$\|Q\|$	[20 *]	Fixed length of the queue of *ad*

* Best hyperparameter values.

The values highlighted in Table 1 were obtained by training the model multiple times to identify the optimal configuration. The best hyperparameters were defined as those yielding the highest accuracy on the validation set. Consequently, these hyperparameters were utilized for training the final model, ensuring enhanced performance based on the established criteria. Conversely, Table 2 shows the model selection for the LSTM model, which was used as the basis for the benchmark model for AD.

Table 2. LSTM model selection.

Model	Hyperparameter	Value	Description
LSTM	hidden_state	[100, 200 *, 300]	Hidden state size
	n_layers	[1, 2 *, 3]	Number of recurrent layers
	dropout	[0 *, 0.5]	Dropout regularization hyperparam.
	lr	0.01 *	Learning rate
	b1	0.9 *	B1 Adam optimizer hyperparam.
	b2	0.99 *	B2 Adam optimizer hyperparam.
	weight_decay	0 *	L2 regularization hyperparam.
	ϵ	[0.8, 1.0 *, 1.2]	ϵ constant of anomaly detector (*ad*)
	$\|Q\|$	[20 *]	Fixed length of the queue of *ad*

* Best hyperparameter values.

3. Results

The first stage of the experiment tested an LSTM model, a widely used approach to AD. The model was trained over 50 epochs with a batch size of 16, using the Adam algorithm for optimization and the MSE as the loss function. In the next stage, the proposed solution, based on the implementation of the ESN model, was applied. With this approach, a rigorous comparison was made between the ESN model paired with an AD block and the current state-of-the-art LSTM networks. Table 3 presents the results of the evaluation, where the TR_Time and TS_Time metrics measure the time (in seconds) required to train the network and perform inference on data. The emissions generated during the training period are expressed in terms of CO_2 and calculated in kg. In order to demonstrate the effectiveness of the proposed solution compared to alternative approaches, a targeted selection of the most relevant features was performed. Among the 88 variables examined, the focus remained on harmonic distortion, with an in-depth analysis of the third harmonic of the current. This variable reflects the degree to which the load current waveform deviates from its ideal sinusoidal shape and is a key parameter not only for assessing the quality of the electrical power but also for detecting anomalies within the system.

Table 3. Comparison between ESN-based and LSTM-based models.

	ESN	LSTM
TR_Time	9.5 s	2100 s
TR Emissions	3.8515×10^{-5}	0.0085
TS_Time	4.47 s	7.39 s
TS Emissions	8.697×10^{-5}	0.0086
TS Accuracy	85%	80.88%
TS Precision	97.17%	96.33%
TS Recall	75.15%	67.96%
TS F1 Score	86.25%	79.83%

Table 3 shows that the ESN-based model performed almost as well as the more complex LSTM-based model while also improving other indicators. Indeed, the TR_Time of the ESN-based model was significantly lower (9.5 s) than that of the LSTM-based model (2100 s), thus effectively reducing the CO_2 emissions (3.8515×10^{-5} compared to 0.0085). Referring to the efficacy metrics, although the TS_Precision of the ESN-based model (97.17%) and that of the LSTM-based model were similar (96.33%), the ESN-based model achieved a higher recall (75.15% compared to 67.96%), indicating that the system was more effective at identifying anomalies without missing them. The same result applied to the accuracy metrics, where the TS_Accuracy of the ESN-based model (85%) surpassed that of the LSTM-based model (80.88%). As previously described, a Raspberry Pi 4 with only 1 GB of RAM served as the experimental edge platform connected directly to the production machine. The results, as shown in Table 3, indicate that the accuracy levels achieved by the LSTM model over 50 epochs were exceeded by the ESN model trained in one shot. The training times demonstrate that the LSTM model required 2100 s, while the ESN model completed training in just 9.5 s, highlighting the significant computational efficiency of the proposed framework during the training phase. Furthermore, the results also reveal that the testing times for the ESN model (4.47 s) were superior to those of the LSTM model (7.39 s), underscoring its enhanced computational performance. Figure 8 shows graphs comparing the error variation and the trend of the time series of the third harmonic of the current. In particular, the following elements are analyzed:

- The measure of the standard deviation of the error indicated by q.

- Anomaly threshold: To ensure that a fault does not occur, the values of the standard deviation must remain within an upper limit ($+\sigma$) and a lower limit ($-\sigma$). The threshold is calculated as the average of the two limits.
- Effective anomaly corresponds to the actual anomaly that can be detected in the system.
- Predicted anomaly corresponds to the prediction of the anomaly by the model.

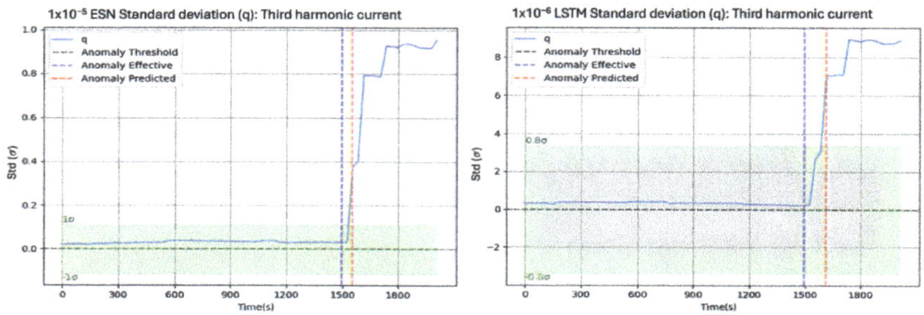

Figure 8. Standard deviation of the error of both the ESN and LSTM model-based approaches.

On the left, the ESN-based model graph indicates the predicted anomaly occurrences with dashed red lines and the effective anomalies with solid blue lines. The green shaded region depicts the confidence interval, specifically $\pm 1\sigma$ from the mean. Anomalies are detected when the standard deviation exceeds the anomaly threshold, represented by the dashed green line. Similarly, the right graph depicts the LSTM-based model's performance with analogous markings. However, the LSTM-based model's confidence interval appears broader, as shown by the green shaded area, which may suggest a higher variability in the predicted values during the training phase and the σ computation phase. Notably, the ESN-based model promptly identifies the anomaly at the moment when the effective anomaly surpasses the threshold. In contrast, the LSTM-based model exhibits a delayed response, signaling the anomaly only after a lag, which may indicate a lower recall rate. The ESN-based model's sharper threshold crossing and immediate detection response highlight its agility and potentially higher recall in recognizing deviations from the norm. This performance is indicative of the ESN-based model's suitability for real-time anomaly detection applications at the edge, where timely identification is critical. The plots in Figure 9 show a comparison of the performance of the ESN-based model and LSTM-based model in predicting and detecting anomalies based on the analysis of the third harmonic current over time. The following elements are analyzed:

- Time-Series Effective: This corresponds to the time-series data representing the actual trend.
- Time-Series Prediction: This corresponds to the time-series data representing the predicted trend.
- Effective Anomaly: This corresponds to the beginning of the actual anomaly.
- Predicted Anomaly: This corresponds to the beginning of the predicted anomaly.

Each graph contains two continuous lines representing the actual time-series values (Time-Series Effective) and the predicted time-series values (Time-Series Prediction) in blue and orange, respectively. Each plot in Figure 9 shows two continuous lines representing the actual time-series values (Time-Series Effective) and the predicted time-series values (Time-Series Prediction) in blue and orange, respectively. The left plot illustrates the ESN-based model's performance, where a close alignment between the predicted and effective time series can be observed until the occurrence of an anomaly. However, the ESN-based model exhibits a more consistent and stable prediction and anomaly detection capability, as indicated by the close alignment of the predicted values with the effective time series. In contrast, the LSTM model shows greater fluctuations in its predictions, suggesting that it may be more sensitive to noise in the data, while the anomaly detection appears to lag behind the actual anomalies, which leads to delayed responses in a real-world scenario. Conversely, Figure 10 illustrates the relationship between the number of epochs and the accuracy achieved on the test set, comparing the performance of the LSTM-based method with that of the ESN-based method.

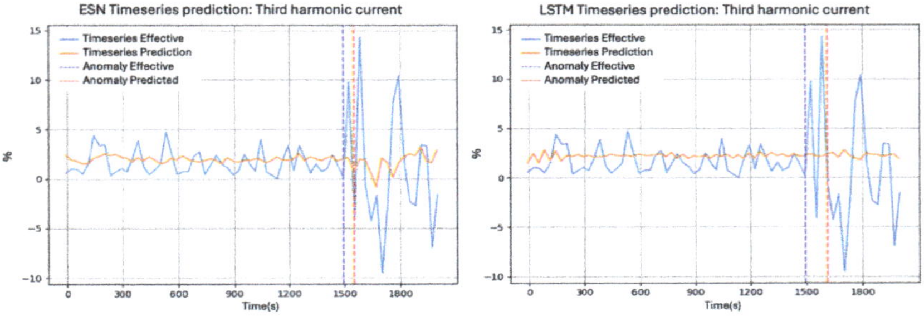

Figure 9. Comparison between real time series and time-series prediction (ESN-based model on the left and LSTM-based model on the right).

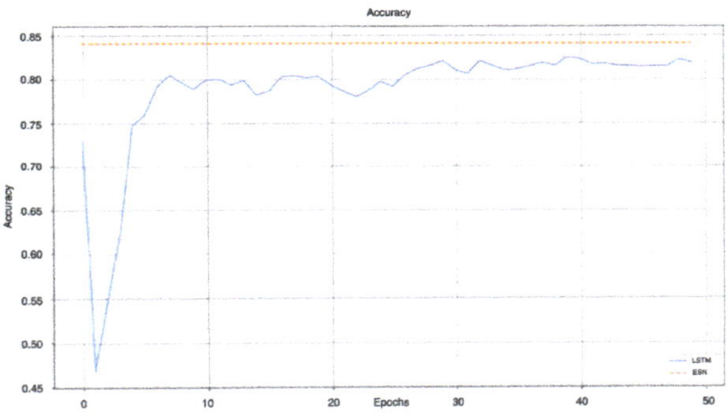

Figure 10. Accuracy for each epoch (LSTM-based and ESN-based methods).

In this analysis, the accuracy of the LSTM method was monitored after each epoch of training and compared to the accuracy achieved by the ESN method, which was trained in one shot. It is evident that the ESN method trained in one shot outperformed the LSTM method trained for 50 epochs in terms of accuracy.

4. Discussion

This section summarizes the observations and discusses the reasons why the proposed model achieved better results and outperformed the baseline methods. The proposed ESN-based lightweight framework primarily consists of an ESN model and an anomaly detector. The ESN model performs next-step predictions by calculating the Mean Squared Error (MSE) between the predicted values and the target values, generating error values for collection in an analysis buffer. Subsequently, the standard deviation of the stored errors is computed. This designed neural network structure effectively achieves a favorable trade-off between reducing computational complexity and retaining critical features for anomaly detection. This result is due to its unique reservoir computing framework, which is particularly effective at processing time-series data and making next-step predictions, making it well suited for analyzing deviations between current and predicted values. Moreover, considering that the foundational structure of the ESN consists of a randomly initialized reservoir and a readout layer whose weights are computed in one shot using ridge regression, it is evident that the proposed framework requires low computational resources, making it particularly suitable for implementation at the edge, close to the system being analyzed. In contrast, the incorporation of other recurrent networks, such as RNNs, LSTM networks, or GRUs, within the proposed framework necessitates training using backpropagation through time, thereby increasing computational resource demands and making it more suitable for server- or cloud-based applications.

In order to validate the proposed solution, tests were conducted by replacing the ESN-based model within the framework with the LSTM-based model. The choice to use the latter as a benchmark can be summarized as follows:

1. The LSTM-based model is one of the most widely used models in industrial anomaly detection applications.
2. Although there are many other valid models in the literature, such as RNNs, GRUs, and Transformer-based architectures, the LSTM model represents a trade-off between high performance, the ability to capture very complex relationships and long-term dependencies, and the computational resources required.
3. The proposed solution was applied to a real industrial scenario using a smart meter with over 250 features (of which only 88 were utilized). The GRU and LSTM models would have faced the high input dimensionality challenge, which means that as the number of input feature variables increases, the performance of the network tends to decrease.

Even though the "Results" section highlights that the proposed solution achieved superior performance compared to the reference benchmark (85 compared to 80.88 s) and required 99.65% fewer computational resources (4.5 s compared to 2100 s), there are several considerations to address in order to make the system more robust and reliable for industrializing the framework. These include the crucial analysis of the relationship between the products manufactured and the trends in electrical parameters. For example, when the material used by the production machine is aluminum, a softer material, the current and voltage values differ compared to when steel, a harder material, is utilized. This complicates the generalization of the model without integration with enterprise information systems that provide data along with information about the items in production. In addition, acquiring longer time-series data and labeling each individual fault or issue presents a significant challenge because it is important not only to identify and classify a diverse range of faults but also to ensure that these labels are accurately aligned with the corresponding timestamps in the time series. Moreover, the reduction in the performance of production machines complicates this process further. As machines experience wear and tear or operational inefficiencies, their output could introduce variable patterns within the time-series data. These variations must be carefully distinguished from real fault conditions to avoid mislabeling and to maintain the integrity of the dataset.

5. Conclusions

This study investigated the feasibility of performing AD on edge devices. This result was achieved using a highly efficient neural network model, namely the Echo State Network (ESN). The findings indicate that the ESN-based approach not only outperforms the traditional LSTM method but is also faster and less resource-intensive, which makes it particularly suitable for use in embedded applications. The comparative performance evaluation highlights the ESN approach as an efficient alternative to the LSTM approach for anomaly detection tasks, especially in scenarios where model stability and training speed are desirable. The higher initial accuracy of the ESN approach suggests that it has an advantage in rapidly understanding and adapting to the structure within the data, which is critical for real-time or near-real-time anomaly detection systems. The successful integration of the proposed framework with a real production machine underscores its practical applicability and relevance in enhancing production efficiency in a real industrial scenario. As the demand for robust anomaly detection continues to grow, the insights gained from this study lay the foundation for further exploration and implementation of ESN-based solutions in different and more complex industrial applications. Moreover, factors such as the influence of the item being produced, the analysis of acquisition frequency, the number of electrical parameters, and the influence of the network's initialization parameters should be studied. Future work will include building a pipeline that introduces continuous learning techniques to mitigate real-world system degradation over time, which is a significant limitation of static techniques, and integrating production planning information with power consumption data in order to increase next-step prediction accuracy.

Author Contributions: Methodology, A.B., M.P., G.P. and L.L.; validation, A.B., M.P., G.P. and L.L.; formal analysis, A.B. and M.P.; investigation, M.P., G.P. and L.L.; resources, M.P., R.K., G.P., L.L. and C.V.; data curation, G.P., R.K., L.L. and C.V.; writing—original draft preparation, M.P., G.P., A.B. and L.L.; writing—review and editing, A.B. and M.P.; visualization, A.B. and M.P.; supervision, S.L. and A.B.; project administration, A.B. and M.P. All authors have read and agreed to the published version of the manuscript.

Funding: This research was supported by Italian PON "RICERCA E INNOVAZIONE" 2014–2020—AZIONE IV.6 "CONTRATTI DI RICERCA SU TEMATICHE GREEN".

Data Availability Statement: The original contributions presented in the study are included in the article, further inquiries can be directed to the corresponding author.

Acknowledgments: We express our deepest appreciation to Sifim Srl for their consent to install IoT sensors on their production machines, and we also extend our gratitude to Syncode Scarl for the hardware devices and cloud service for data storage and visualization.

Conflicts of Interest: Author Carlo Verdini was employed by the company Syncode Scarl. The remaining authors declare that the research was conducted in the absence of any commercial or financial relationships that could be construed as a potential conflict of interest.

Abbreviations

The following abbreviations are used in this manuscript:

AD	Anomaly Detection
AI	Artificial Intelligence
BPTT	Backpropagation Through Time
DT	Digital Twin
ESN	Echo State Network
GAN	Generative Adversarial Network
GAE	Graph AutoEncoder
IIoT	Industrial Internet of Things
LoRaWAN	Long-Range Wide Area Network
LSTM	Long Short-Term Memory
MAE	Mean Absolute Error
MSE	Mean Squared Error
MPC	Model Predictive Control
ML	Machine Learning
RNN	Recurrent Neural Network
SD	Standard Deviation
THD	Total Harmonic Distortion
VAE	Variational AutoEncoder

References

1. Shang, C.; You, F. Data analytics and machine learning for smart process manufacturing: Recent advances and perspectives in the big data era. *Engineering* **2019**, *5*, 1010–1016. [CrossRef]
2. Zhai, Q.; Cao, H.; Zhao, X.; Yuan, C. Cost benefit analysis of using clean energy supplies to reduce greenhouse gas emissions of global automotive manufacturing. *Energies* **2011**, *4*, 1478–1494. [CrossRef]
3. Domingo, R.; García, M.; Sánchez, A.; Gómez, R. A sustainable evaluation of drilling parameters for PEEK-GF30. *Materials.* **2013**, *6*, 5907–5922. [CrossRef] [PubMed]
4. Zou, J.; Chang, Q.; Ou, X.; Arinez, J.; Xiao, G. Resilient adaptive control based on renewal particle swarm optimization to improve production system energy efficiency. *Int. J. Manuf. Syst.* **2019**, *50*, 135–145. [CrossRef]

5. Bonci, A.; Longhi, S.; Nabissi, G. Fault Diagnosis in a belt-drive system under non-stationary conditions. An industrial case study In Proceedings of the 2021 IEEE Workshop on Electrical Machines Design, Control and Diagnosis (WEMDCD), Modena, Italy, 8–9 April 2021; pp. 260–265.
6. Muthiah, K.M.N.; Huang, S.H. Overall throughput effectiveness (OTE) metric for factory-level performance monitoring and bottleneck detection. *Int. J. Prod. Res.* **2007**, 4753–4769. [CrossRef]
7. Braglia, M.; Castellano, D.; Frosolini, M.; Gallo, M. Overall material usage effectiveness (OME): A structured indicator to measure the effective material usage within manufacturing processes. *J. Prod. Plan. Control* **2018**, 143–157. [CrossRef]
8. Busogi, M.; Ransikarbum, K.; Oh, Y.G.; Kim, N. Computational modelling of manufacturing choice complexity in a mixed-model assembly line. *Int. J. Prod. Res.* **2017**, 5976–5990. [CrossRef]
9. Bonci, A.; Stadnicka, D.; Longhi, S. The Overall Labour Effectiveness to Improve Competitiveness and Productivity in Human-Centered Manufacturing. In *International Scientific-Technical Conference MANUFACTURING*; Lecture Notes in Mechanical Engineering; Springer: Cham, Switzerland, 2022; pp. 144–155.
10. Prist, M.; Monteriù, A.; Pallotta, E.; Cicconi, P.; Freddi, A.; Giuggioloni, F.; Caizer, C.; Verdini, C.; Longhi, S. Cyber-physical manufacturing systems: An architecture for sensor integration, production line simulation and cloud services. *IEEE J. Acta IMEKO* **2020**, *4*, 39–52. [CrossRef]
11. Prist, M.; Longhi, S.; Monteriù, A.; Giuggioloni, F.; Freddi, A. An integrated simulation environment for Wireless Sensor Networks. In Proceedings of the IEEE 16th International Symposium on A World of Wireless, Mobile and Multimedia Networks (WoWMoM), Boston, MA, USA, 14–17 June 2015; Volume 4, pp. 1–5.
12. Grisostomi, M.; Ciabattoni, L.; Prist, M.; Ippoliti, G.; Longhi, S. Application of a wireless sensor networks and Web2Py architecture for factory line production monitoring. In Proceedings of the IEEE 11th International Multi-Conference on Systems, Signals & Devices (SSD14), Castelldefels-Barcelona, Spain, 11–14 February 2014; pp. 1–6.
13. Zenisek, J.; Holzinger, F.; Affenzeller, M. Machine learning based concept drift detection for predictive maintenance. *Comput. Ind. Eng.* **2019**, *137*, 106031. [CrossRef]
14. Tanuska, P.; Spendla, L.; Kebisek, M.; Duris, R.; Stremy, M. Smart Anomaly Detection and Prediction for Assembly Process Maintenance in Compliance with Industry 4.0. *Sensors* **2021**, *21*, 2376. [CrossRef]
15. Tancredi, G.P.; Vignali, G.; Bottani, E. Integration of Digital Twin, Machine-Learning and Industry 4.0 Tools for Anomaly Detection: An Application to a Food Plant. *Sensors* **2022**, *22*, 4143. [CrossRef] [PubMed]
16. Li, H.; Li, Y. Anomaly detection methods based on GAN: A survey. *Appl. Intell.* **2022**, *53*, 8209–8231. [CrossRef]
17. Tagawa, Y.; Maskeliūnas, R.; Damaševičius, R. Acoustic Anomaly Detection of Mechanical Failures in Noisy Real-Life Factory Environments. *Electronics* **2021**, *10*, 2329. [CrossRef]
18. Pol, A.A.; Berger, V.; Germain, C.; Cerminara, G.; Pierini, M. Anomaly Detection with Conditional Variational Autoencoders. In Proceedings of the 18th IEEE International Conference On Machine Learning And Applications (ICMLA), Boca Raton, FL, USA, 16–19 December 2019; Volume 53, pp. 1651–1657.
19. Han, P.; Ellefsen, A.L.; Li, G.; Holmeset, F.T.; Zhang, H. Fault detection with lstm-based variational autoencoder for maritime components. *IEEE Sens. J.* **2021**, *21*, 21903–21912. [CrossRef]
20. Kim, H.; Lee, B.S.; Shin, W.-Y.; Lim, S. Graph Anomaly Detection with Graph Neural Networks: Current Status and Challenges. *IEEE Access* **2019**, *10*, 111820–111829. [CrossRef]
21. Fan, G.; Ma, Y.; Mei, X.; Fan, F.; Huang, J.; Ma, J. Hyperspectral anomaly detection with robust graph autoencoders. In Proceedings of the IEEE International Conference on Acoustics, Speech and Signal Processing ICASSP, Seoul, Republic of Korea, 14–19 April 2024; Volume 21, pp. 1830–1834.
22. Malhotra, P.; Vig, L.; Shroff, G.; Agarwal, P. Long Short-Term Memory Networks for Anomaly Detection in Time Series. In Proceedings of the 23rd European Symposium on ANN, Bruges, Belgium, 22–24 April 2015.
23. Terzi, M.B. Anomaly Detection with Deep Long Short Term Memory Networks. In Proceedings of the 6th International Conference on Computer Science and Engineering (UBMK), Ankara, Turkey, 15–17 September 2021; Volume 21, pp. 129–132.
24. Li, W.; Ng, W.; Wang, T.; Pelillo, M.; Kwong, S. HELP: An LSTM-based approach to hyperparameter exploration in neural network learning. *Neurocomputing* **2021**, *442*, 161–172. [CrossRef]
25. Nguyen, V.Q.; Van Ma, L.; Kim, J. LSTM-based anomaly detection on big data for smart factory monitoring. *J. Digit. Contents Soc.* **2018**, *19*, 789–799.
26. Mustafa, A.; Joung, B.G.; Lee, W.J.; Mousoulis, C.; Raghunathan, N.; Shakouri, A.; Sutherland, J.; Bagchi, S. Anomaly detection and inter-sensor transfer learning on smart manufacturing datasets. *Sensors* **2023**, *19*, 486. [CrossRef]
27. Wang, L.; Lv, S.X.; Zeng, Y.R. Effective sparse adaboost method with ESN and FOA for industrial electricity consumption forecasting in China. *Energy* **2018**, *155*, 1013–1031. [CrossRef]
28. Pan, Y.; Wang, Y. Model predictive control of unknown nonlinear dynamical systems based on recurrent neural networks. *IEEE Trans. Industr. Electron.* **2011**, *59*, 3089–3101. [CrossRef]
29. Matino, I.; Dettori, S.; Colla, V.; Weber, V.; Salame, S. Forecasting blast furnace gas production and demand through echo state neural network-based models: Pave the way to off-gas optimized management. *Appl. Energy* **2019**, *253*, 113578. [CrossRef]
30. Miao, D.; Feng, K.; Xiao, Y.; Li, Z.; Gao, J. Gas Turbine Anomaly Detection under Time-Varying Operation Conditions Based on Spectra Alignment and Self-Adaptive Normalization. *Sensors* **2024**, *24*, 941. [CrossRef] [PubMed]

31. Bono, F.M.; Radicioni, L.; Cinquemani, S.; Bombaci, G. A Comparison of Deep Learning Algorithms for Anomaly Detection in Discrete Mechanical Systems. *Appl. Sci.* **2023**, *13*, 5683. [CrossRef]
32. Do, J.S.; Kareem, A.B.; Hur, J.-W. LSTM-Autoencoder for Vibration Anomaly Detection in Vertical Carousel Storage and Retrieval System (VCSRS). *Sensors* **2023**, *23*, 1009. [CrossRef] [PubMed]
33. Panza, M.A.; Pota, M.; Esposito, M. Anomaly Detection Methods for Industrial Applications: A Comparative Study. *Electronics* **2023**, *12*, 3971. [CrossRef]
34. Tiboni, M.; Remino, C.; Bussola, R.; Amici, C. A Review on Vibration-Based Condition Monitoring of Rotating Machinery. *Appl. Sci.* **2022**, *12*, 972. [CrossRef]
35. Dastjerdi, A.V.; Buyya, R. Fog Computing: Helping the Internet of Things Realize. *IEEE Comput. Soc.* **2016**, *49*, 112–116. [CrossRef]
36. Lindemann, B.; Maschler, B.; Sahlab, N.; Weyrich, M. A survey on anomaly detection for technical systems using LSTM networks. *Comput. Ind.* **2021**, *131*, 10349. [CrossRef]
37. Braei, M.; Wagner, S. Anomaly Detection in Univariate Time-Series: A Survey on the State of the Art. *arXiv* **2020**, arXiv:2004.00433.
38. Bagus, B.; Gepperth, A. An Investigation of Replay-based Approaches for Continual Learning. In Proceedings of the International Joint Conference on Neural Networks (IJCNN), Gold Coast, Australia, 18–23 June 2021; pp. 1–9.
39. Grigoryeva, L.; Ortega, J.P. Echo state networks are universal. *Neural Netw.* **2018**, *108*, 495–508. [CrossRef]
40. Jaeger, H. *The "echo state" Approach to Analysing and Training Recurrent Neural Networks*; German National Research Center for Information Technology GMD Technical Report: Bonn, Germany, 2001.
41. Verstraeten, D.; Schrauwen, B.; D'Haene, M.; Stroobandt, D. An experimental unification of reservoir computing methods. *Neural Netw.* **2007**, *20*, 391–403. [CrossRef] [PubMed]
42. Hochreiter, S.; Schmidhuber, J. Long short-term memory. *Neural Comput.* **1997**, *9*, 1735–1780. [CrossRef] [PubMed]
43. Fenza, G.; Gallo, M.; Loia, V. Drift-Aware Methodology for Anomaly Detection in Smart Grid. *IEEE Access* **2019**, *7*, 9645–9657. [CrossRef]
44. Bonci, A.; Longhi, S.; Nabissi, G.; Scala, G.A. Execution Time of Optimal Controls in Hard Real Time, a Minimal Execution Time Solution for Nonlinear SDRE. *IEEE Access*, **2020**, *8*, 158008–158025. [CrossRef]
45. Carvalho, A.; Machado, C.; Moraes, F. Raspberry Pi Performance Analysis in Real-Time Applications with the RT-Preempt Patch. In Proceedings of the 2019 Latin American Robotics Symposium (LARS), 2019 Brazilian Symposium on Robotics (SBR) and 2019 Workshop on Robotics in Education (WRE), Rio Grande, Brazil, 23–25 October 2019; Volume 442, pp. 162–167.
46. Longhi, M.; Taylor, Z.; Popović, M.; Nieto, J.; Marrocco, G.; Siegwart, R. RFID-Based Localization for Greenhouses Monitoring Using MAVs. In Proceedings of the 2018 IEEE-APS Topical Conference on Antennas and Propagation in Wireless Communications (APWC), Cartagena, Colombia, 10–14 September 2018; pp. 905–908.
47. Smart Meter Seneca S604. Available online: Https://www.seneca.it/media/4166/power_1912eng_r1.pdf (accessed on 4 March 2024).
48. Tiwari, D.; Miscandlon, J.; Tiwari, A.; Jewell, G.W. A Review of Circular Economy Research for Electric Motors and the Role of Industry 4.0 Technologies. *Sustainability* **2021**, *13*, 9668. [CrossRef]
49. Seeniselvi, T.; Nirmala, M. A survey on data preparation and feature engineering in machine learning. *Int. J. Adv. Res. Comput. Eng. Technol. (IJARCET)* **2019**, *8*, 191–198.

machines

MDPI

Article

PolyDexFrame: Deep Reinforcement Learning-Based Pick-and-Place of Objects in Clutter

Muhammad Babar Imtiaz *, Yuansong Qiao and Brian Lee

Software Research Institute, Technological University of the Shannon, Midlands Midwest, N37 HD68 Athlone, Ireland
* Correspondence: muhammad.babarimtiaz@tus.ie

Abstract: This research study represents a polydexterous deep reinforcement learning-based pick-and-place framework for industrial clutter scenarios. In the proposed framework, the agent tends to learn the pick-and-place of regularly and irregularly shaped objects in clutter by using the sequential combination of prehensile and non-prehensile robotic manipulations involving different robotic grippers in a completely self-supervised manner. The problem was tackled as a reinforcement learning problem; after the Markov decision process (MDP) was designed, the off-policy model-free Q-learning algorithm was deployed using deep Q-networks as a Q-function approximator. Four distinct robotic manipulations, i.e., grasp from the prehensile manipulation category and inward slide, outward slide, and suction grip from the non-prehensile manipulation category were considered as actions. The Q-function comprised four fully convolutional networks (FCN) corresponding to each action based on memory-efficient DenseNet-121 variants outputting pixel-wise maps of action-values jointly trained via the pixel-wise parametrization technique. Rewards were awarded according to the status of the action performed, and backpropagation was conducted accordingly for the FCN generating the maximum Q-value. The results showed that the agent learned the sequential combination of the polydexterous prehensile and non-prehensile manipulations, where the non-prehensile manipulations increased the possibility of prehensile manipulations. We achieved promising results in comparison to the baselines, differently designed variants, and density-based testing clutter.

Keywords: polydexterous; deep reinforcement learning; prehensile; non-prehensile; robotic manipulation; Markov decision process; deep Q-network; fully convolutional network; pixelwise-parameterization; DenseNet-121

1. Introduction

The concept of Industry 4.0 has been prevalent for over a decade now [1]. The fourth industrial revolution, centered on automation, introduced key technologies such as Digital Twins, Cyber–Physical Systems, Smart Manufacturing, and the Internet of Things (IoT). These innovations have become integral to implementing automated processes within industries. With this revolution of industry, one can clearly see the direction and future of industries, as described by Warren Bennis [2]: "The factory of the future will have only two employees, a man and a dog. The man will be there to feed the dog. The dog will be there to keep the man from touching the equipment." The proof-of-concept of the Industry 4.0 revolution can be witnessed all over the world, for example, Europe's Industry 4.0, China's Made in China 2025, USA's Advanced Manufacturing, Japan's Super Smart Society 5.0, etc. One can witness different terminologies such as smart factories, smart manufacturing, and smart industry in the literature, which fall under the umbrella of Industry 4.0. In these domains, while achieving the goal of automation, the involvement of artificial intelligence (AI) and the utilization of robotics are quite evident.

Robotics has been a part of industry for almost five decades, evolving in various forms over time [3]. The operations usually performed by industrial robots can be divided into three categories: material handling, processing operation, and assembly and inspection [4]. Tasks included in the material handling category are material transfer and machine loading and unloading. In processing operations, robotic arms are used to perform a process on a medium such as welding and painting. In assembling and inspection operations, products are assembled, and their quality is ensured. In all these categories, the most frequent and repetitive robotic manipulation is the pick-and-place operation. In addition to pure industrial robots, even collaborative robots, also known as cobots, are specially designed to work alongside human workers, mostly to perform pick-and-place tasks. Over the years, along with pure industrial robots, collaborative robots have become the focus of attention of industry due to their ability to quickly learn on the job and trouble-free reprogramming features. It is being witnessed that with the rapid growth of industries and e-commerce, pick-and-place or bin-picking robotic arms are in demand to achieve automation in various factories and warehouses [5].

Usually, robotic manipulations are categorized into two categories: prehensile manipulation [6,7] and non-prehensile manipulation [8]. The key difference between the two is that the former implies grasping or capturing and the latter lacks the element of grasping and instead adopts pushing or propelling. Figure 1 clearly showcases the difference between both categories. Our previous work [9] showed that a well-coordinated approach between these two types of robotic manipulations can significantly enhance success rates. Therein, we found that non-prehensile manipulation such as sliding can navigate through tightly packed objects, facilitating subsequent grasping. Meanwhile, prehensile manipulation such as grasping can efficiently displace objects, ensuring collision-free and precise movements for non-prehensile manipulations, such as sliding or pushing. Previously, much research was conducted separately on prehensile and non-prehensile robotic manipulations. However, there remains a substantial research gap and ample opportunity to explore the integration of these two types of robotic manipulations that we targeted in our previous work [9].

Figure 1. The difference between prehensile and non-prehensile manipulations.

When we consider the combination of prehensile and non-prehensile robotic manipulation, the role of non-prehensile manipulation can vary, for example, modifying the object poses, creating gaps between objects, breaking down a clutter of objects for further manipulation, etc. In most existing research combining prehensile and non-prehensile robotic manipulation, the varying role of non-prehensile manipulation has not been thoroughly examined, neither in model-based work [10] nor in data-driven approaches [11]. The primary difference between the model-based and data-driven is that the former relies entirely on static models, whereas the latter involves models trained through learning from relevant data. In prehensile manipulation, some approaches learn from previous experiences [12] while others rely on grasp stability measures [13]. However, most current approaches lack the ability to sequentially learn a combination of prehensile and non-prehensile manipulations, which could enhance the success rate of pick-and-place tasks.

In this research study, we extend our previous work [9] and explore the possibility of integrating previous work on the agent learning synchronization of grasping and sliding with suction-based grasping, which technically falls under the category of non-prehensile manipulation. To extend the previous work, we develop a polydexterous framework, which uses different robotic grippers on a robotic arm to perform multiple robotic manipulations. The major objectives or contributions of this research study are as follows:

- A deep reinforcement learning-based end-to-end framework, known as PolyDexFrame, enables the agent to learn the sequential combination of jaw-gripper and suction cup/pad prehensile and non-prehensile robotic manipulations, thus allowing the pick-and-place of regular- and irregular-shaped objects successfully.
- The jaw-gripper grasping, bidirectional sliding, and suction-cup-based vacuum gripping are learned together in such a sequential manner where each robotic manipulation further enhances the chances of the other manipulations.
- Optimal policy convergence is achieved through Q-learning by training joint end-to-end CNNs in a self-supervised mode with the help of a pixelwise-parameterization technique.

An important question here regards the need or reason behind extending our research by integrating the suction-based non-prehensile manipulation with the jaw-gripper-based prehensile and non-prehensile manipulations. There are several scenarios in industry and warehouses where there is clutter to be picked and placed, comprising various kind of objects; some may be deformable, soft, and flexible objects, whereas others may be rigid ones. When a regular jaw gripper faces problems in accurately grasping the deformable, soft objects, the suction cup/pad can be the right option due to its gentleness, adaptability to the soft deformable object's surface, and reduced pressure points [14]. Therefore, the agents' learning to combine the prehensile and non-prehensile manipulations of different grippers, i.e., to be polydexterous, in a sequential manner which may lead to an enhanced success rate of the pick-and-place task is highly beneficial. Therefore, the novelty of this extended research work is evident in the joint learning of the jaw-gripper grasping, bidirectional sliding, and suction-cup-based vacuum gripping for the pick-and-place of regular- and irregular-shaped clutter.

2. Related Work

In this research study, we target the amalgamation of robotic gripper-based grasping, sliding, and vacuum gripping with the help of a suction cup/pad. Therefore, we examine these three elements addressed individually thus far in the existing body of literature.

The robotic gripper-based grasping has been studied and practiced for a long time now. A lot of the literature consists of various model-based approaches. Some of these approaches identified the merit of individual grasps based on their ability to control the movement of the object [15], whereas a couple of them explored the resistive role of contact forces by modeling them in the process of grasping [16]. In some approaches, the main idea is to relate point clouds with the grasps computed beforehand dynamically on the run for the sake of the pose estimation of the objects under consideration [17]. The pre-computed grasps are mostly generated with the help of 3D object models which make the real-world adaptation much better [18]. The obvious limitation of these approaches is that a large amount of in-depth information is necessary beforehand for them to work, such as the shape of the objects, poses of the objects, pre-computed contact points on the objects, etc. But when the objects are novel and unknown, then it is impossible to meet this mandatory condition, which makes these approaches unsuitable. On the other hand, researchers have been working on data-driven approaches instead in recent years. Various data-driven approaches are evident, in which agents are trained in a model-free manner to identify the potential grasps after learning the visual features instead of requiring any in-depth information beforehand [12]. Approaches like [19] present the idea of using models pretrained to learn other robotic manipulations to be used for the learning of the desired robotic manipulation. The approach in [12] is completely deep-learning-based and tends to calculate pixelwise affordances in order to make the model learn the optimal grasping policy.

Like the history of prehensible manipulation, the relevant literature suggests that non-prehensile manipulation has also been around for a long time. The early approaches developed in the non-prehensile domain used the assistance of frictional forces in modeling [20]. Even though these approaches played a vital role in the future development of the non-prehensile research area, the modeling assumptions made during the process, such as non-uniform distribution and varying frictional forces, proved to be insufficient for real-world settings [21]. These limitations causing hurdles in adapting to real-world settings led to the exploration of data-driven approaches for the learning of non-prehensile manipulations [22]. Most of the techniques presented in these approaches revolved around dealing with a single object. However, exploring dense clutter remains a complex challenge in this domain.

The vacuum gripping with the help of a suction cup/pad is one of the widely deployed non-prehensile manipulations in warehouses and industries [23,24]. In the three popular events of the Amazon Picking Challenge [25], most approaches were based on vacuum gripping using a suction cup/pad. The Amazon Picking Challenge 2015 featured more than half of its approaches using suction grippers, including the winning team [25], which achieved suction gripping by pushing objects to the sides and walls. In [26], the authors utilized 6D pose estimation to perform vacuum gripping using a suction gripper near the center of the object surface. In [27], both 6D models and manual heuristics were employed for suction gripping at potential grasp coordinates. Fully convolutional networks were used by [12] to link visual inputs to the probabilities of suction grasps. In [28], the authors used two different masks to achieve grasp estimation from a single depth image. These masks represented the contact and collision regions, respectively. Another approach, [29], involved training convolutional neural networks to predict the grasp quality by classifying potential suction grasps in cloud points. A dataset comprising 2.8 million point clouds, suction grasps, and relevant labels was created for this purpose. Additionally, double-stream convolutional neural networks have been used to calculate suction point probabilities in clutter, with the aid of manual annotations [30]. Despite these advancements, most deep-learning-based approaches still rely on data manually annotated by humans.

The research area combining prehensile and non-prehensile robotic manipulations has not been fully explored yet, but there is no doubt about its efficacy. The study presented in [8] can be seen as a pioneering model-based approach that tried to combine prehensile and non-prehensile manipulations. It performs the combination of push–grasp, where the grasping is achieved while the pushing through clutter, with the hope of increasing the chances of grasping. However, this approach requires a large amount of relevant in-depth information beforehand, whereas the proposed approach in this research study requires absolutely no information beforehand since it learns in a self-supervised mode. Some other combination-based approaches [11] have also been presented where the grasping pose and locations are predefined, and the role of the non-prehensile pushing is to bring the objects to these pre-designed grasp locations. In these approaches, in addition to the limitation of a static grasping part, the pushing part is also not learned from scratch. A reinforcement-learning-based approach [31] has also been presented using handcrafted features to make the agent learn and select from prehensile and non-prehensile manipulations to pick-and-place the clutter. In this approach, object detection relies on the segmentation of RGB-D images, after which potential manipulations/actions are sampled for each detected object, and for each sampled action, handcrafted features are extracted before performing that manipulation. The downside of this approach is that it is only suitable for clutter composed mostly of convex objects. Another limitation of the approach is the requirement for prior information regarding the pushing actions, such as motion prediction through a simulator. Using the reinforcement learning, some singulation techniques [32] have also been presented, restricted to regular shapes and limited non-prehensile manipulation.

In the research area of combining prehensile and non-prehensile robotic manipulations, little has been carried out in combining suction-based gripping non-prehensile manipulation with jaw-gripper-based prehensile and non-prehensile manipulations. The proposed approach in this paper addresses this research gap. This work is essentially an extension of the research conducted in [9], which combined jaw-gripper-based prehensile and non-prehensile manipulations, and it extends the work in [7], where the pixelwise-parametrization technique was adopted for prehensile manipulation. The role of Q-learning in this work prior to adopting the pixelwise-parameterization technique was presented in [33,34].

3. Background

In this section, we briefly review the basic concepts of the cutting-edge technologies of deep learning, reinforcement learning, and their amalgamation, known as deep reinforcement learning, involved in this end-to-end data-driven approach.

3.1. Deep Learning

Deep learning is a very well-known member of the machine-learning family because of its exceptional performance in various domains. Deep learning is not limited to computer vision problems, its application can be found in many fields, including climate science, medical science, bioinformatics, drug design, material inspection, machine translation, natural language processing, speech recognition, social network filtering, etc. Deep learning has even outperformed human experts in board games. In other words, deep learning has introduced the world to many new ways of data processing, analysis, and manipulation. The word "deep" in deep learning refers to the presence of multiple layers in a single network. Whereas in machine learning, feature extraction from input data is performed manually by experts, deep learning extracts higher-level features automatically as the input data pass through the multiple layers of the deep neural network, as shown in Figure 2. Here, input data refers to raw images. Usually, large volumes of data are required in deep learning to achieve maximum efficiency and optimal results. Because it deals with large volumes of data, it requires substantial computational power. Deep models can be seen as artificial neural networks involving deep structures. The idea of the artificial neural network can be traced back to the early 1940s [35]. Since then, several turning points and landmarks have been achieved, with the introduction of perceptron, backpropagation method, rectified linear unit, max-pooling, dropout, batch normalization, etc. This series of algorithmic advances, the availability of large-scale data, and the emergence of highly efficient parallel computing entities such as Graphics Processing Units (GPUs) have contributed to the success of deep learning today. The first major achievement in the field of deep learning was achieved by a convolutional neural network (CNN) performing classification after training in 2012 [36]. It was trained on hundreds of thousands of labelled images, using loss computation and backpropagation to learn the parameters. Since then, this method has seen many improvements and can be considered the most popular deep learning approach; however, it is not well-suited for learning robotic manipulations. The primary reason for its unsuitability for learning robotic manipulation is its very high time requirement, which necessitates a large number of labelled images of joint angles for training. Nevertheless, researchers are exploring alternatives [37], and deep reinforcement learning, which will be discussed later, appears to be more efficient and promising.

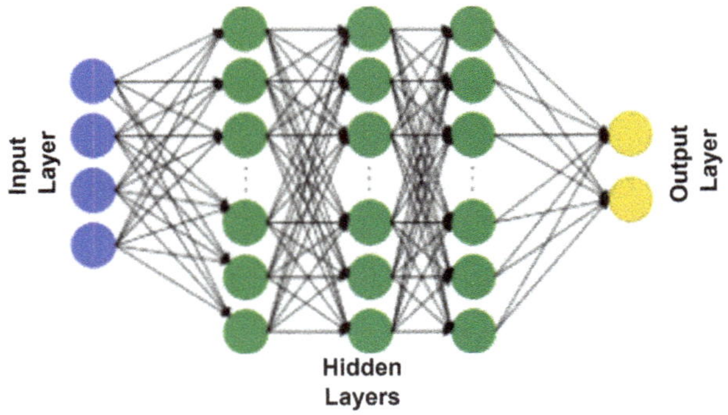

Figure 2. The flow diagram of a deep neural network.

3.2. Reinforcement Learning

Reinforcement learning is one of the subdomains of the machine-learning field [38]. It focuses on finding behaviors that maximize rewards through a process of trial and error. This mirrors how humans and animals learn: by perceiving and processing information, storing it as knowledge, and adapting their behavior in response to changing circumstances. This self-supervised technique is being deployed in several fields such as statistics, information theory, system optimization, recommendations systems, operations research, game theory, and control theory [39]. Natural reinforcement learning can be seen in living beings through punishment or reward, thus reinforcing positive behavior and suppressing negative behavior. In the same manner, an agent starts in a state, takes an action according to its current policy, receives a reward from the environment, ends up in a new state, chooses a new action, and continues iterating through this procedure. For a better understanding, this cycle is shown in Figure 3. Simply put, it is like playing a video game in which a character performs actions to maximize their score. Reinforcement learning is not like supervised learning, where labeled datasets are available for training the agent. When deploying reinforcement learning in interactive use cases, such as robotic control problems, labeled data or depictions of all potential actions are often scarce. Therefore, the concept of rewards is introduced. Reinforcement learning is also not like unsupervised learning, where we identify patterns or structures in unlabeled datasets. Instead of finding patterns and structures, the agent gradually learns to behave in the environment based on future rewards. Therefore, it can be termed as the third paradigm of machine learning, next to supervised and unsupervised learning [40].

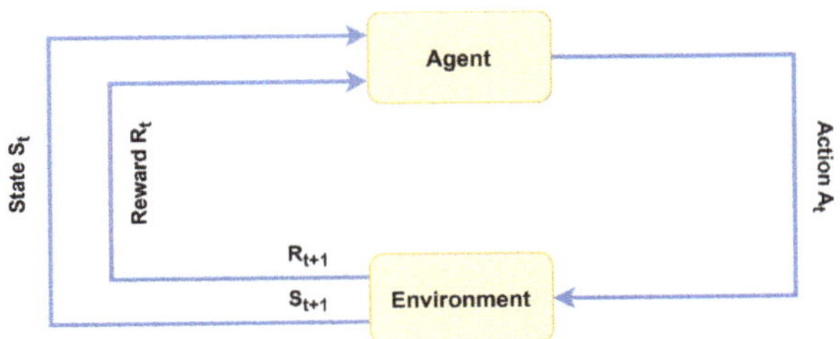

Figure 3. The working of a reinforcement-learning-based agent.

3.3. Deep Reinforcment Learning

As evident from the name, deep reinforcement learning is an amalgamation of reinforcement learning and deep learning. From reinforcement learning, it borrows the technique of awarding rewards to agents based on the actions they chose to perform, and from deep learning, it uses the idea of learning features through the deployment of neural networks. While traditional reinforcement learning is suited for environments with simple state representations, deep reinforcement learning enables agents to learn from unstructured and multi-dimensional inputs using neural networks. A general outlook of a deep reinforcement learning-based approach is shown in Figure 4. In the last few years, as deep learning has gained popularity, deep reinforcement learning has also made significant progress in various domains. Deep reinforcement learning gained much of its popularity through its role in games research. In 2013, DeepMind produced significant results in the Atari game without relying on hand-coded features. Later, in 2016, AlphaGo made headlines by defeating a human expert in the ancient Chinese game Go. Its improved version, AlphaZero, produced in 2017, demonstrated proficiency in learning features in shogi and chess as well. In 2019, Pluribus and OpenAI Five outperformed top professionals in multiplayer poker and defeated the defending world champion in Dota 2, respectively. Beyond games, deep reinforcement learning holds potential in other domains such as robotics, computer vision, natural language processing, finance, healthcare, and transportation. Many tech giants, academic labs, and non-profit research organizations have focused on applying deep reinforcement learning and have published several breakthroughs. The widely used deep reinforcement learning algorithms can generally be classified into three categories: value-based, policy gradient, and model-based methods. In the first category, value-based methods, the policy is defined by a value function based on an algorithm using the Bellman equation, such as Q-learning or its variant, fitted Q-learning. The successful implementation of the deep Q-network (DQN) and its various extensions, including double DQN and distributional DQN, are examples of this category. The performance of these methods, in terms of final results and data efficiency, is benchmarked against the Atari 2600 environment in [41]. Deep Q-network (DQN) approaches handle problems with discrete action spaces and deterministic policies. On the other hand, policy gradient methods address problems with continuous action spaces and stochastic policies by finding a policy based on neural network parameters to increase the expected cumulative reward. Policy gradient methods require an estimate of a value function for the current policy, like other policy methods, and deploy actor–critic architectures through a sample-efficient approach, enabling the use of off-policy data. An example of such a method is the Deep Deterministic Policy Gradient (DDPG). Research on combining Q-learning and policy gradient methods is also ongoing. Both value-based and policy-based methods face limitations in sample efficiency because they do not utilize the environmental model, making them model-free approaches. In contrast, model-based methods learn an environmental model through experience, using approximation functions or pre-supplied models, in conjunction with a planning algorithm. Studies have also explored combining model-based and model-free methods to leverage the strengths of both approaches.

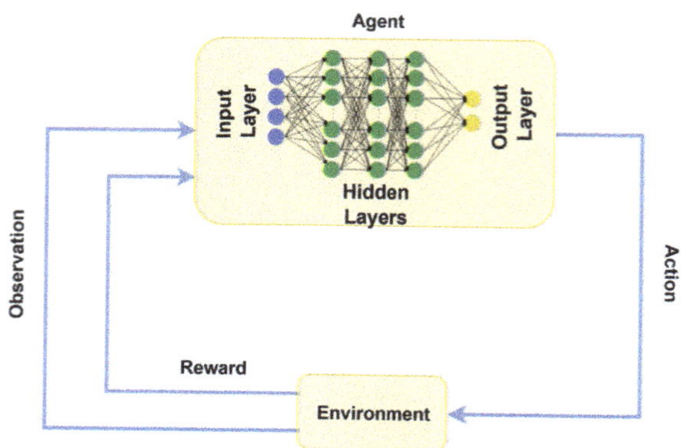

Figure 4. The block diagram represents the working of a deep reinforcement learning-based agent.

4. Methodology

In this research study, we develop a polydexterous robotic pick-and-place framework. The reason behind terming it a polydexterous framework is that "poly" means many and "dexterous" means skill with the hand. Thus, a polydexterous robotic arm can be seen as a robotic arm performing prehensile and non-prehensile manipulations while involving more than one robotic gripper, which is the case in this proposed approach.

In this research paper, we design a pick-and-place framework for variously shaped objects, including regular and irregular ones. The task is modeled as a deep reinforcement learning problem, where the agent tries to learn the skill of the pick-and-place of regular- and irregular-shaped objects from a conveyor belt through a continuous sequential combination of prehensile and non-prehensile robotic manipulations with the help of two robotic grippers. We train an agent through reinforcement learning to learn the pick-and-place task. An agent here means an abstract entity that iteratively learns a policy to maximize its expected reward, thereby mastering the task. As the task is to be resolved as a reinforcement learning problem, we formulate a Markov Decision Process (MDP), which is a quintuple process comprising states, actions, transition function, rewards, and a discount factor. Thus, according to the MDP flow, an agent (the robotic arm in our case) at any time t, residing in the initial state s_t, takes an action a_t while following the policy π, and ends up in the next state s_{t+1}. For this transition from the initial state to the next state, it also earns some reward, denoted by $r_{at}(s_t, s_{t+1})$. Thus, in terms of MDP, the actual goal here for the agent is to find the optimal policy π^* which can lead the agent to reach the goal of maximum reward, depicted by $R_t = \Sigma_{i=t}^{\infty} \gamma r_{ai}(s_i, s_{i+1})$. Here, the γ represents the discount factor, valued between 0 and 1, which helps the agent to learn with more speed.

This approach is essentially a self-learning, data-driven scheme where the agent (the UR5 robotic arm we are deploying) tries to predict the most suitable pixel or coordinate for the relevant robotic manipulation to be performed. As soon as the agent becomes successful in predicting the accurate pixel for the manipulation, the overall pick-and-place process by becoming proficient in the sequential synchronization of the prehensile and non-prehensile manipulations also becomes successful, as evident from the results. In this approach, we deployed the off-policy, model-free Q-learning algorithm to enable the agent to learn the pick-and-place task through prehensile and non-prehensile robotic manipulations. The Q-learning agent tends to learn the optimal policy in a greedy manner, always choosing the action with the maximum action-value. These action-values are calculated through the Q-function, which can be depicted as $Q_\pi(s_t, a_t)$. The working of a Q-function can be summed up as in Equation (1).

$$Q(s_t, a_t) \leftarrow Q(s_t, a_t) + \alpha\left[R_{at}(s_t, s_{t+1}) + \gamma Q\left(s_{t+1}, argmax\left(Q\left(s_{t+1}, a_t'\right)\right) - Q(s_t, a_t)\right)\right] \quad (1)$$

where the α and a_t' represent the learning rate and the directory of all the action options available at the time.

To understand the flow of the approach, the main three elements of the MDP are described as follows:

- **States:** A state ($\mathbf{s} \in \mathbf{S}$) is the current snapshot of the agent's environment. In our case, states are represented graphically. Each state is depicted as an RGB-D heightmap, which shows the graphical view of the agent's environment. As the UR5 robotic arm is our agent, the conveyor belt containing the regular and irregular objects is seen in its surrounding environment at any given time. This surrounding environment can also be called the workspace of our agent within which it can observe and perform actions. Since the state is designed graphically, vision sensors are needed to capture the current state of the UR5 robotic arm's workspace. We installed two kinds of vision sensors: an orthographic vision sensor and a perspective vision sensor. These sensors have different fields of view: the orthographic vision sensor's field of view is rectangular, while the perspective vision sensor's field of view is trapezoidal. The reason for using these two different vision sensors is to install them at fixed positions, as shown in

Figure 5, and then merge their field of views to collect the maximum amount of RGB and depth information from the environment. Both vision sensors capture their own individual RGB-D images, which are later amalgamated into one image. As the name suggests, the RGB-D image contains the color information (R, G, and B channels) as well as the depth information (D channel). The depth information represents the distance of objects' surfaces from a viewpoint. Since our robotic manipulation is from the top angle, the depth data can be considered elevation or height-from-bottom data. As stated above, the state is represented in RGB-D heightmap format. The RGB-D image is translated to the corresponding RGB-D heightmap using the same scheme as we used before in [9]. As shown in Figure 6, after generating and back-projecting the 3D point cloud from the input RGB-D data in an orthographic manner, the RGB-D heightmap is procured. The size of the generated RGB-D heightmap corresponds to the actual workspace of the agent, which is equivalent to 224^2 pixels in our case. By this design, each of the 50,176 pixels in the RGB-D heightmap corresponds to a distinct 3D location in the physical workspace of the agent designed in the simulator.

- **Actions**: An action ($a \in A$) is a step chosen from the available list performed by the agent after observing the environment, following the policy to maximize its expected future reward. In our case, the actions are the prehensile and non-prehensile robotic manipulations performed by the UR5 robotic arm agent to pick-and-place regular and irregular objects in clutter. We consider four different robotic manipulations: grasp, inward-slide, outward-slide, suction-grip. These actions cover both prehensile and non-prehensile robotic manipulation. In our setup, two robotic grippers are involved: the jaw gripper and the suction cup. The grasp is the prehensile manipulation of the jaw gripper, while the inward-slide and outward-slide are its non-prehensile manipulations, as shown in Figure 7. The suction-grip is the non-prehensile robotic manipulation performed by the suction cup, as shown in Figure 7. In summary, there is one prehensile manipulation and three non-prehensile manipulations that the agent can learn to synchronize such that each robotic manipulation increases the chances of successfully picking and placing objects.
At any given time, the agent can select a pixel from the 224^2 sized RGB-D heightmap. As discussed before, each pixel corresponds to a particular 3D location on the robotic arm's workspace on the conveyor belt represented as $pixel \rightarrow 3D\ location \in S_t$. This 3D location on the workspace has different roles depending on the action. For instance, in the case of grasping action, it will be considered the mid-point of the grasp. In the inward- or outward-slide, it will be the start point of the linear push. In the suction-grip, it is the contact point. According to the design setup, some static modifications are made to the runtime when performing the action in the given 3D location. For instance, because we have vision sensors installed at top angles and the pick-and-place is performed from the top angle, the z-coordinate of the predicted 3D location on the workspace is incremented by around 2 cm in the cases of grasp and suction-grip. The increment ensures that the object can be grasped from center point, and in the case of suction gripping it helps to create vacuum suction pressure inside the suction cup, leading to a stronger grip. In non-prehensile manipulations, the direction of the slide is towards the robotic arm base in the inward-slide and away from the robotic arm base in the outward-slide. For these sliding non-prehensile manipulations, the straight linear push limit is kept at 8 cm.
- **Rewards**: The reward ($r \in R$) in the reinforcement learning is the incentive that the agent receives when it transitions from one state to another by choosing the appropriate action. The goal of the agent is to maximize its expected future reward. In our case, we have a reward scheme consisting of discrete reward values. For a successful grasp and placement, the agent is awarded the full reward, which is 1. If the grasp is successful but the placement fails, meaning the object slips from the gripper before placement, the reward is reduced to 0.8. To determine whether the object has slipped from the jaw gripper, a ray-type infrared proximity sensor is installed between

the jaws of the gripper. For successful inward- or outward-slide actions, the reward is 0.5, with success measured by comparing the workspace before and after sliding. If changes are detected, the action is considered successful, and a reward is awarded. The suction-grip action has the highest chance of success due to the lower probability of slipping, and therefore it is rewarded lightly at 0.2. In any scenario other than the ones stated above, the agent is rewarded 0. This reward scheme encourages the agent to prioritize actions that lead to the successful grasping and placing of objects while also providing feedback for partial success and less optimal actions.

- **Q-Function:** In reinforcement learning, the Q-function generates action-values representing the expected future reward an agent will receive if it takes a specific action in a given state and transitions to the next state. In simple problems, this can be maintained as a Q-table, but in more complex problems, such as ours, neural networks are used for function approximation. Since we are deploying an off-policy model-free Q-learning algorithm and using deep neural networks, this approach is referred to as deep Q-networks (DQNs) for transition function approximation [42,43]. In our approach, we design the Q-function using multiple fully convolutional networks (FCN) [44]. The goal is to enable the agent to learn the synchronization among the four distinct robotic manipulations to enhance the success rate of the picking and placing variously shaped objects in clutter. To achieve this, we deploy a distinct FCN for each robotic manipulation, allowing the agent to learn the important features of all the robotic prehensile and non-prehensile manipulations in substantial depth. We designed four FCNs: FCN_{Grasp}, $FCN_{Inward\text{-}Slide}$, $FCN_{Outward\text{-}Slide}$, and $FCN_{Suction\text{-}Grip}$. Each of these FCNs receives the state as input, represented as an RGB-D heightmap of the workspace, sized as 224^2 pixels. All four FCNs are fed the same RGB-D heightmap simultaneously at any given time. This approach uses the pixelwise-parameterization technique, where processing is performed at the pixel level. Consequently, the output generated by each FCN is a 224^2-sized pixelwise map of values. The input and output are the same size because of this technique. Thus, we supply the same RGB-D heightmap to all four FCNs and receive four distinct pixelwise maps of the same size. The values in these pixelwise maps are the action-values on which the agent bases its decision for the current state. Each value in the pixelwise map predicts the expected future reward if the agent performs the particular action at the corresponding 3D location of the predicted pixel in the workspace. This setup allows the agent to evaluate and choose actions that maximize its expected reward, improving its effectiveness in performing pick-and-place tasks.

All four distinct FCNs share the same neural architecture, specifically DenseNet-121 [45]. The reason for adopting the DenseNet architecture is due to the connections between all layers. In a traditional FCN with n number of layers, each layer is only connected to its next layer, resulting in n number of connections. However, in DenseNet, a network with n layers has $n(n + 1)/2$ total connections, meaning that each layer receives features from all preceding layers and transmits features to all succeeding layers. This enhances feature reuse and propagation, reduces the vanishing gradient problem, and minimizes the number of parameters. Despite these advantages, the increased connectivity and depth can lead to an exponential quadratic growth of the features, causing a bottleneck. To address this vulnerability, we use a memory efficient implementation of DenseNet which will be detailed in the next section. DenseNet-121 was trained and tested in two settings: from scratch and pretrained on ImageNet [36]. The DenseNet-121 architecture was extended by adding two pointwise convolution layers and batch normalization operations, with ReLU as the activation function. Our approach employs a multimodal architecture as shown in Figure 8. Each FCN has two separate trunks: one for color data (RGB) and one for depth data (DDD). The depth channel, originally a single, is cloned to match the three color channels using the method used in [9]. Later, these trunks merge, passing their learned features into the extended pointwise convolutional layers.

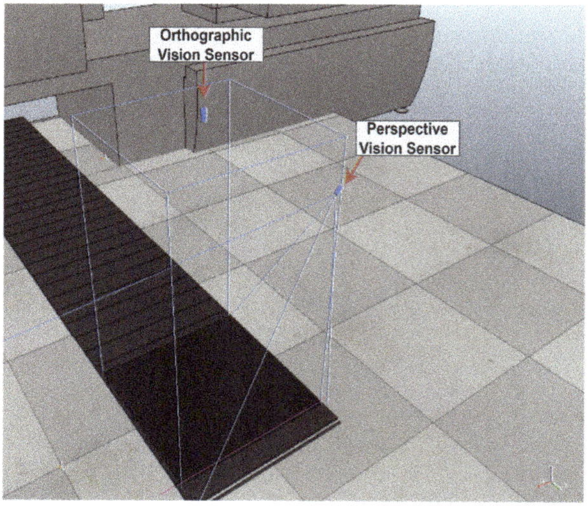

Figure 5. Orthographic and perspective vision sensors' field of view.

Figure 6. Generation of RGB-D heightmap from the RGB and depth components.

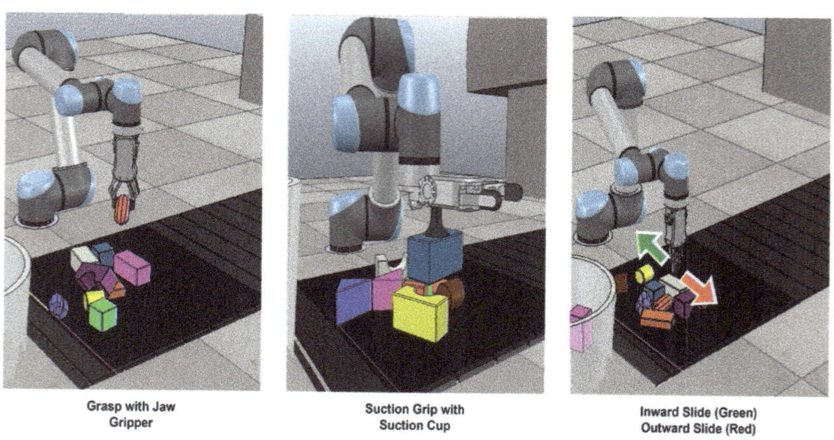

Figure 7. Examples of robotic arm performing grasping, suction-gripping, inward-slide, and outward-slide.

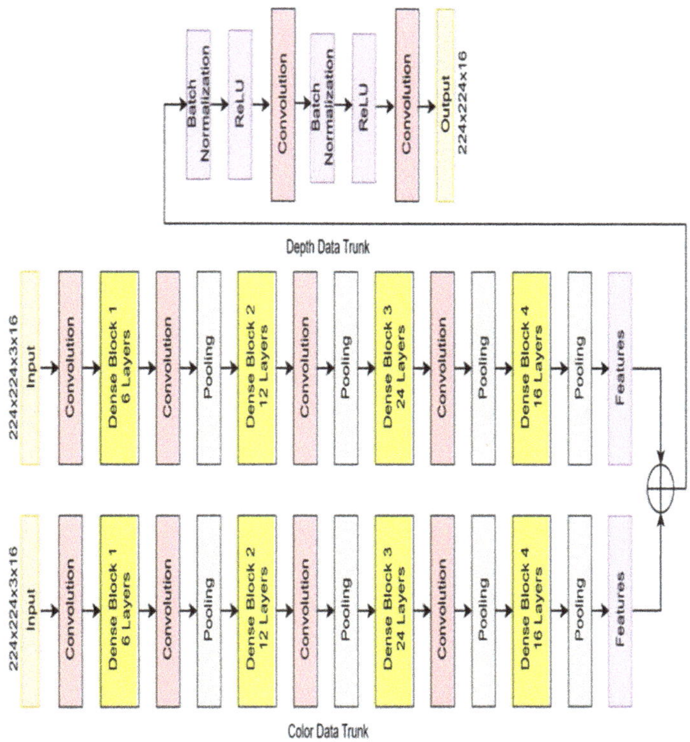

Figure 8. The block diagram of the multimodal extended DenseNet-121.

In our previous work, we have employed a scheme of rotations of the current state view to enhance feature learning. We apply the same rotation technique in this case too. To compare the performance, we rotated the RGB-D heightmap by 90°, 45°, and 22.5° as shown in Figure 9. These rotation settings generate 4, 8, and 16 rotations, respectively, for each RGB-D heightmap state. Similar to our previous results, the maximum feature learning was noted at the rotation setting of 22.5°. Adopting the 16-rotation setting means that each of the FCNs is supplied with 16 rotated versions of the RGB-D image, resulting in 16 pixelwise maps generated by each FCN. In one single iteration of the FCNs, 64 pixelwise maps of action-values are produced, with 16 pixelwise maps for each action: grasp from the FCN_{Grasp}, inward-slide from the $FCN_{Inward\text{-}Slide}$, outward-slide from the $FCN_{Outward\text{-}Slide}$, and the suction-grip from the $FCN_{Suction\text{-}Grip}$. Each pixelwise map contains 50,176 (224^2) action-values, resulting in a total of 802,816 (50,176 × 16) action-values for each action. The decision of which action to perform out of four is made by the agent by selecting the maximum action-value from all 3,211,264 (50,176 × 16 × 4) action-values combined. In other words, the corresponding action is performed at the 3D location corresponding to the pixel predicted with the maximum action-value/Q-value out of all the 64 pixelwise Q-values maps, as expressed in Equation (2).

$$\mathbf{max}Q(s_t, a_t) = \mathbf{max}\left(FCN_{grasp}(s_t),\ FCN_{Inward-Slide}(s_t), FCN_{Outward-Slide}(s_t), FCN_{Suction-Grip}(s_t)\right) \quad (2)$$

With the help of the pixelwise-parameterization technique along with the rotation scheme described above, our approach enables the agent to learn at an optimal learning rate with a minimal amount of training data required.

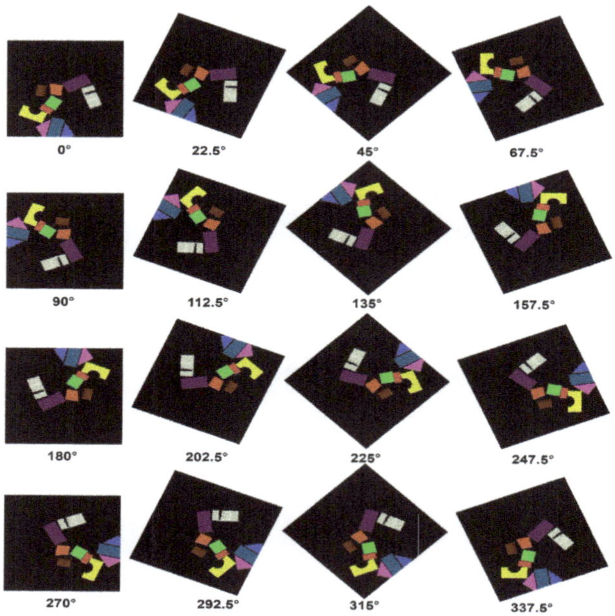

Figure 9. RGB-D heightmap rotations by 22.5°.

4.1. Training Specifications

In this section, we describe the process step by step to provide a better view of the training cycle, as shown in Figure 10. The cycle starts with regular and irregular objects entering the workspace on the conveyor belt. A ray-type infrared proximity sensor installed at the workspace detects the objects, and the vision sensors capture the current situation of the workspace in the form of RGB-D images, which are processed to generate the RGB-D heightmap. This RGB-D heightmap is passed to the rotation scheme function, which takes the RGB-D heightmap as the input, performs rotation at 22.5°, and produces 16 rotations of the RGB-D heightmap. These 16 rotations are fed to all four FCNs, and they generate the 16 pixelwise maps of action-values/Q-values for the agent. To represent these 3,211,264 (50,176 $\times$ 16 $\times$ 4) action-values, we use a heatmap representation as shown in Figure 11, which can compactly sum up the huge amount of data in less space and highlight the higher Q-values by showing them as hot regions in orange–red shades. In Figure 10, the approach is shown as a top-down tree structure. The RGB-D heightmaps from color and depth data act as the root of the tree, and the next level shows four nodes of the tree representing the 16 pixelwise maps of Q-values presented in heatmap format. The leaves of this tree show the 3D locations corresponding to the predicted pixels with the highest Q-values on the workspace (represented by the red dots). Each red dot represents the highest Q-value 3D location for the corresponding action to be performed. At this stage, the Q-values of all these four 3D locations/pixels are compared, and the one with the maximum value is chosen to perform the corresponding action.

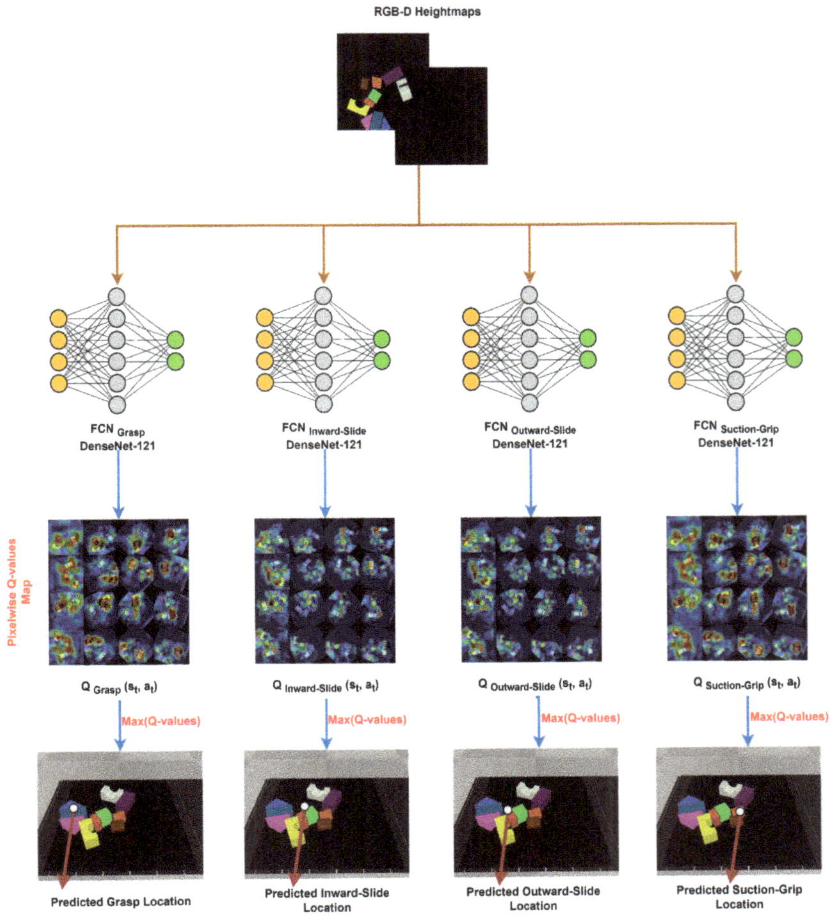

Figure 10. The flowchart diagram of the proposed approach.

As we discussed for the forward pass, backpropagation is also crucial for the weight tuning. To leverage the benefits of two widely used loss functions, i.e., mean square error (MSE) and mean absolute error (MAE), we deployed the Huber Loss function [46]. During the backpropagation, gradient passing is performed only for the FCN that predicts the pixel with the maximum Q-value. For the remaining pixels, zero loss is backpropagated. For the optimizer, the stochastic gradient descent with momentum (SGDM) variant [47] is used in this approach instead of basic SGD. The reason behind choosing SGDM over SGD is its ability to provide more stable values, being closer to the actual loss due to exponential weighted averages. This helps the agent to obtain faster convergence. In the training hyperparameters, the momentum is kept at 0.9, the weight decay is set to 1×10^{-3}, and the learning rate is set to 1×10^{-2}. An epsilon value is defined to address the exploration vs. exploitation dilemma and it is kept between 0.1 and 0.4, with a gradual decrease. In terms of hardware specifications, our system is equipped with an Intel® Core™ i7-9700K processor, 16 GB DDR3 RAM, and a GeForce RTX 2080 GPU with 8GB memory capacity. For the development, training, and testing of the FCN models, we used PyTorch version 2.2.1.

As discussed before, the reason behind selecting the DenseNet architecture is its design of maximum connections between layers, which leads to maximum feature reuse and propagation while diminishing the vanishing gradient problem. However, this design can make the system vulnerable to exponential quadratic growth of the features. This potentially causes a bottleneck situation that chokes the network. To remedy this vulnerability, we implement a memory-efficient variant of DenseNet-121 [48] instead of the base variant. In this variant of DenseNet-121, a memory sharing scheme is employed for all operations such as batch normalization, concatenation, gradients, etc. Temporary storage buffers are introduced to accommodate the outputs generated after these operations in the network, instead of being allocated new memory every time.

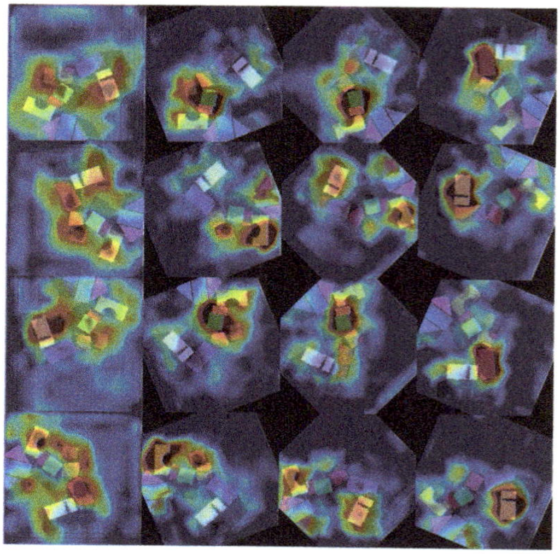

Figure 11. Visualization of 16 pixelwise Q-values maps through heatmap representation.

In the training phase, our agents learn through self-supervision. In each training iteration, the workspace has a certain number (=<10) of regular- and irregular-shaped objects placed in a random clutter pattern, as shown in Figure 12. The agent's task is to clear the workspace by picking and placing all the objects. The iteration is marked complete as soon as the workspace is cleared, and then the next iteration starts with another random clutter of regular- and irregular-shaped objects in the workspace, and the cycle continues.

Figure 12. Regular- and irregular-shaped 3D objects.

4.2. Simulation Setup

For the experimental setup design, we chose the virtual robot experimentation (V-REP) robotic simulator. This simulator not only provides coding options but is also equipped with physics engines that simulate real-world physics effects, such as ODE, Bullet, etc. Multiple different programming platforms, such as Python and C++, can be used with the simulator from outside through threaded and non-threaded Lua scripts inside the simulator. In our case, we implemented the DRL side in PyTorch in Python externally and then communicated with the simulated setup inside the simulator using Lua scripting. Moreover, in addition to the programming support, it also includes various planning modules, such as motion planning and path planning. It features both forward and inverse kinematics modules. The difference between the two is that the forward kinematics module takes all joint angles as the input and provides the end-effector coordinates as the output, while the inverse kinematics module does the opposite. It takes the end-effector position as the input and calculates all the joint angles. In our scenario, we deployed the inverse kinematics module. Another crucial module provided by the simulator is the collision detection and avoidance module, which highlights if any part of the robotic arm collides with the environment. In our designed simulation, we used the popular 6-degree-of-freedom cobot UR5 with two grippers: the RG2 jaw gripper and suction-cup gripper, as shown in Figures 13 and 14. The RG2 jaw gripper handles the grasp, inward-slide, and outward-slide manipulations, while the suction-grip manipulation is performed by the suction-cup gripper. As shown in Figure 15, our simulated experimental setup includes a conveyor belt, on which regular- and irregular-shaped object clutter moves and enters the workspace. The objects are detected by the vision sensors and a proximity infrared sensor, initiating the pick-and-place process. Once the workspace is cleared, the next clutter reaches the workspace, and the cycle continues.

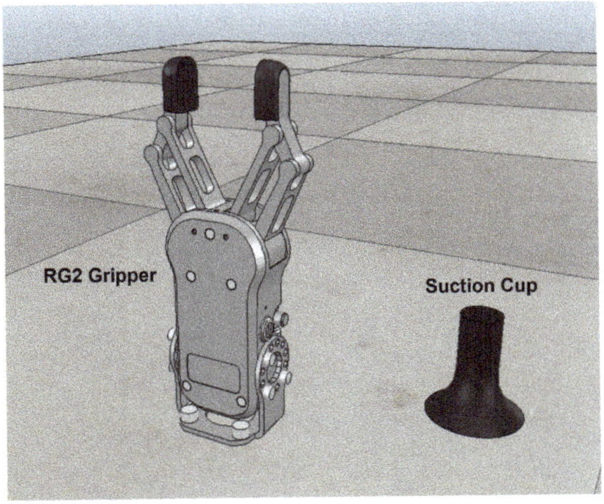

Figure 13. Jaw gripper and suction cup.

To implement the motion planning scheme for the UR5 robotic arm for the pick-and-place task, we explored various options available in the literature, such as OpenRave [49] and Trajpot [50], and finally settled on the Open Motion Planning Library (OMPL) [51]. The reason for this selection was the flexibility it offers for customization and control. OMPL provides several options for control-based and geometry-based planners, and many widely used planning state-spaces are also available. In our work, we deployed a single-query planner known as RRT-Connect [52], which is a bidirectional variant of the traditional RRT. RRT-Connect outperforms traditional RRT because, unlike the base variant, it works with two trees: one at the start and another at the end, which are later connected.

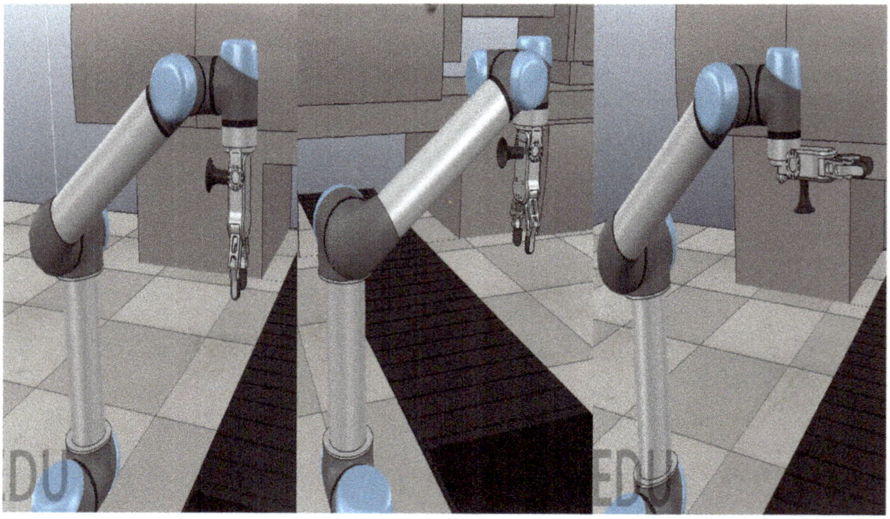

Figure 14. RG2 gripper and suction-cup installation.

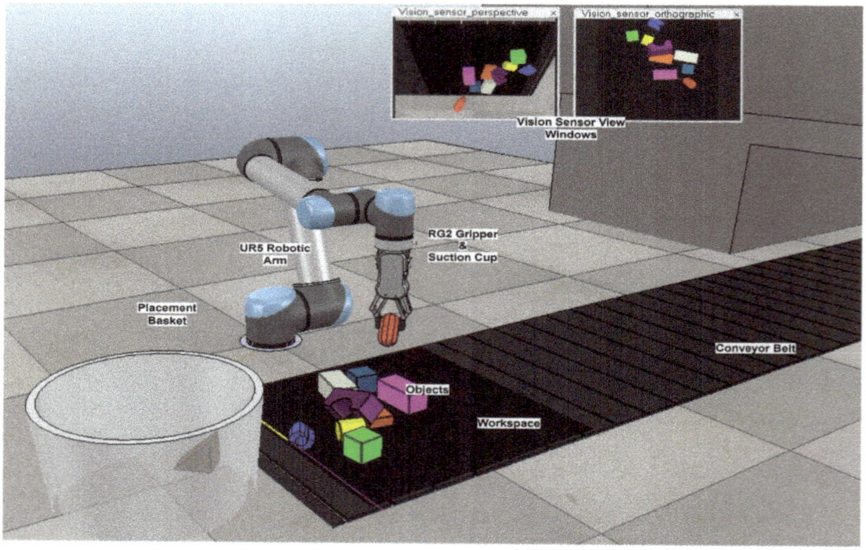

Figure 15. Simulation testbed design.

5. Results and Discussion

This section comprises a series of experiments designed to assess the performance of the developed agents within the simulated environment. The purpose is to evaluate the agents' ability to learn the synchronization of prehensile and non-prehensile manipulations for the pick-and-place task. These experimental results also highlight the role of non-prehensile manipulations in facilitating the prehensile manipulations.

For assessing the performance of our designed agent in terms of pick-and-place, we compare its performance with two different baselines. The first baseline is a variant of the proposed approach that is also based on the pixelwise-parameterization scheme. The major difference is that this baseline lacks the reinforcement learning element, being purely designed as a deep-learning-based binary classification approach. The structural design is the same as the proposed approach, with the representation of states, the design of actions, and the number and architecture of FCNs remaining consistent. While both use the same input and output pixelwise maps of action-values, the difference lies in the agent's training method. In the proposed approach, the agent is trained in self-supervision mode through Q-learning, whereas in this baseline, the FCNs are trained in supervised mode using binary classification. Labels are produced at runtime: a successful action is labeled as 1, and an unsuccessful one as 0. The backpropagation approach is also the same as in the proposed approach. This baseline approach has been previously implemented in [12] with a different architecture and in [9] with the same DenseNet-121 architecture. The comparison of the performance of the proposed approach and the binary classification baseline variant is shown in Figure 16. In this experiment, a series of 3000 episodes is conducted where each episode involves random clutter of ten regular- and irregular-shaped objects being introduced to the workspace. The episode is considered complete when the workspace is cleared or when 10 consecutive failed actions occur. In Figure 16, the orange line represents the proposed approach, while the green line represents the binary classification baseline. The plot shows that the proposed approach outperforms the binary classification, achieving up to an 89% success rate, while the baseline approach does not exceed 70%. The success rate is defined as the number of successfully picked-and-placed objects divided by the total number of actions performed. The lower success rate of the binary classification baseline is likely due to its greedy policy, which prevents it from recognizing long-term strategies that lead to higher expected rewards, unlike the proposed approach.

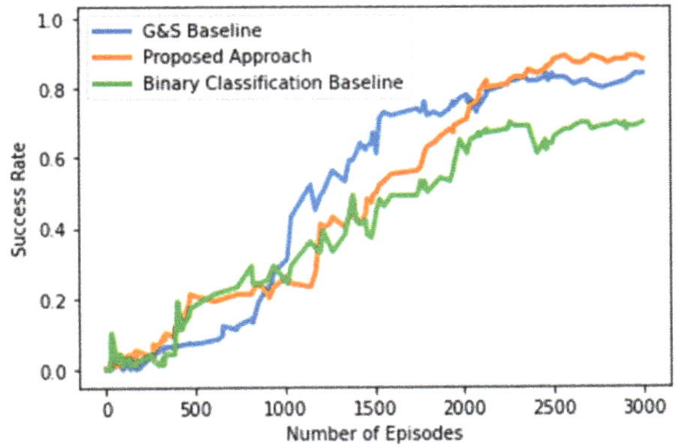

Figure 16. Performance comparison of the proposed approach and the baselines.

The second baseline is deployed from [9], where the agent has only three designed actions: grasping, left-slide, and right-slide. The major difference between the proposed approach and the G&S baseline from [9] is that the latter involves only one robotic gripper and lacks the suction gripping non-prehensile manipulation. The blue line in Figure 16 demonstrates the performance of the G&S baseline against the performance of the proposed approach, represented by the orange line. The addition of the second gripper and the suction gripping action, along with bidirectional sliding and grasping, leads to an improvement of around 5% in the success rate compared to the G&S baseline. This improvement is due to the role of suction gripping in facilitating the other three actions—grasping, inward-slide, and outward-slide—leading to successful pick-and-place operations in the clutter.

To investigate the role of FCNs architecture in the learning of the agent, we designed another variant of the proposed approach. Since we are specifically targeting the role of the neural network architecture, the only difference between the proposed approach and this variant is the architecture of the deployed FCNs. The proposed approach comprised FCNs designed using the DenseNet-121 architecture, but in this variant, the DenseNet-

121 architecture is replaced by the pretrained ResNet-101 [53] architecture. The ResNet architecture consists of residual blocks; hence, it is called a residual network. Any deep neural network has a certain threshold of its depth. As the layers keep on adding and that threshold is breached, the performance deterioration becomes evident because of the widely known vanishing gradient problem. The vanishing gradient problem is basically the extreme shrinking of the gradient while being backpropagated through the depths due to repetitive multiplications. ResNet was presented as the remedy to the vanishing gradient problem. The cure for the vanishing gradient problem in the ResNet architecture lies in the connections known as skip connections. As is evident from the name, the skip connections are named as such because they skip a few layers in gradient transmission, resulting in reduced repetitive multiplications, which leads to no vanishing gradient problem. Figure 17 presents the performance comparison of the proposed approach (in blue) and the ResNet101-based variant (in red). The graph clearly shows that the ResNet101-based variant is beaten by the proposed approach by around a 17% margin. The most probable reason behind this result is the difference in feature maps' handling between the two architectures. The ResNet implements feature map summation, whereas the DenseNet performs feature map concatenation. This results in better feature reusability in DenseNet. Moreover, where the skip connections in the ResNet limit the connectivity among the layers comparatively, the $n(n + 1)/2$ connections in the DenseNet enable higher feature propagation too.

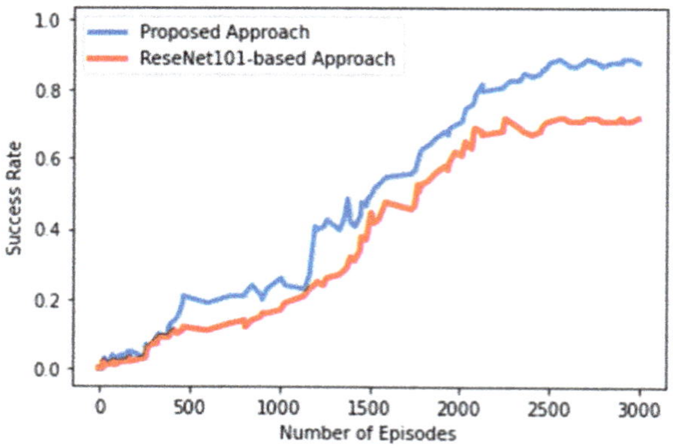

Figure 17. Performance comparison of the proposed approach and the ResNet101-based variant.

In this research study, we explore the role of other non-prehensile manipulations behind the prehensile manipulation. The idea is to evaluate the performance of these non-prehensile manipulations in the convergence of the policy, leading to a high success rate for the prehensile manipulations. To understand the role of non-prehensile manipulations—inward-slide, outward-slide, and suction-grip—we developed a variant of the proposed approach which awards zero rewards for all three non-prehensile manipulations. Rewards were only given for successful prehensile manipulation, i.e., grasping. If the whole pick-and-place was successful, the reward given was 1. If the placement failed after successful grasp, the reward given was 0.8. Figure 18 presents the performance of the no sliding-suction rewards variant (in magenta) against the proposed approach (in blue). The graph shows a fall of around 26% in the performance of the variant compared to the proposed approach, depicting the vital role of non-prehensile manipulations in the convergence of the policy. The delayed rise of the magenta line clearly shows the slower pace of learning of the agent due to the absence of the non-prehensile manipulations' rewards, leading to a grasping-based policy learning. It is possible that the agent is learning an indirect pattern of non-prehensile manipulations leading to prehensile manipulation rewards, but not at the desirable rate.

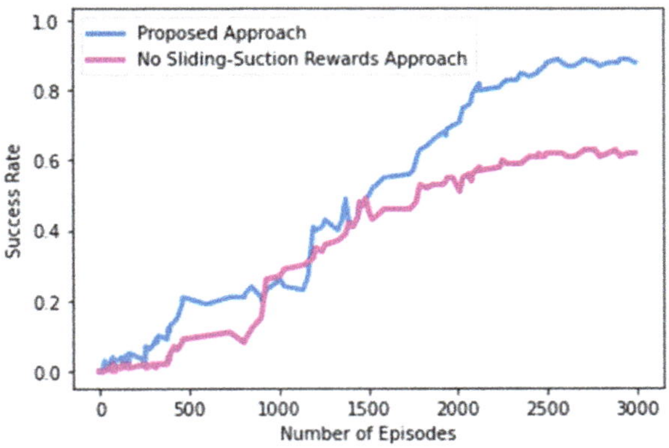

Figure 18. Performance comparison of the proposed approach and the no sliding-suction rewards variant.

In the proposed approach, we deployed the DenseNet-121 architecture pretrained on the ImageNet dataset, which contains a large number of images across approximately 1000 object classes. The general reason behind using pretrained networks is to achieve a better accuracy and reduce training time. In this series of experiments, we also explore the role of pretrained weights in the agent's learning of synchronization between prehensile and non-prehensile manipulations. A variant of the proposed approach was developed with a difference in weights. While the proposed approach uses pretrained weights from the ImageNet dataset, the variant uses no pretrained weights and is trained from scratch. Figure 19 shows the performance of the no pretrained weights variant against the proposed approach. The former is represented by the purple line, whereas the latter is shown by the blue line. According to these results, there is not much difference between the two in terms of performance and success rate. The pretrained weights do not play a significant role in the learning process of the agent. The probable cause behind the lack of benefit from pretrained weights in our case could be the difference in pixel patterns between the RGB-D heightmaps and the RGB images used in the ImageNet dataset. Not only do the pretrained weights seem to have minimal impact on the agent's learning, but the early rise of the purple line also highlights a faster pace of learning. In the case of the proposed approach with pretrained weights, the agent likely needs to first "clean the slate" of the previously learned weights to exit local optima and then start fresh learning, resulting in a delayed rise in the learning curve.

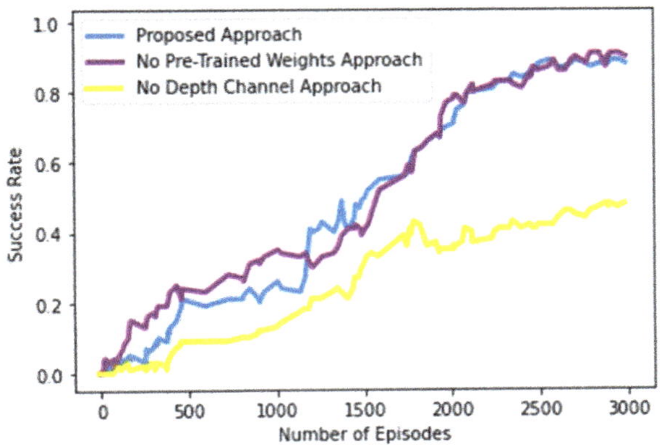

Figure 19. Performance comparison of the proposed approach and the no pretrained weights and no depth channel variants.

We also explore the role of the depth data fed to the FCNs in training the agent. As discussed earlier, in addition to the RGB channels, we utilize the cloned DDD depth channels. To evaluate the role of the depth channel, we designed a variant of the proposed approach that lacks depth channel information. As previously mentioned, in the proposed approach each FCN has two trunks: one is supplied with color information, and the other one is given depth information. In this variant, the major difference is that there is no second trunk designed for depth information. Therefore, feature learning is carried out solely on the color information maps. Figure 19 presents the performance of the no depth channel variant (in yellow) against the performance of the proposed approach (in blue). The results clearly show a significant degradation in performance in the absence of feature learning from the depth data. The lack of height-from-bottom information reduces the performance by approximately 41% compared to the proposed approach. This result underscores the importance of depth data in the agent's learning for the pick-and-place task and its convergence to the optimal policy.

In our training phase, we conducted a total of 3000 episodes, with the number of regular and irregular objects in the random clutter kept at 10. To evaluate the trained model, we used varying clutter sizes. We designed three clutter sizes as shown in Figure 20. The first is the minimum-size clutter, which consists of 6 to 10 randomly distributed regular and irregular objects. The second is called medium-size clutter, containing 15 to 20 regularly and irregularly shaped objects randomly distributed in the workspace. The last is named as maximum-size clutter, where the object count ranges from 25 to 30. The average performance results of the trained agent for these clutter densities are presented in Table 1.

Figure 20. Clutter size categorization [11].

Table 1. Average testing performance.

Clutter Size	Success Rate	Grasp Success	Suction-Grip Success
Minimum-Sized	89%	97%	100%
Medium-Sized	87%	96%	100%
Maximum-Sized	82%	85%	99%

Table 1 presents the average success rate, grasp success, and the suction-grip success of testing across varying densities of clutter. The success rate is defined as the number of successfully picked-and-placed objects divided by the total number of actions performed. Grasp success is calculated as the number of successful grasps in an episode divided by the total number of attempted grasps in that episode. The suction-grip success rate is calculated in the same manner. The overall results indicate that performance is lowest in the maximum-sized clutter. The primary reason for this lower performance in the maximum-sized clutter is the random and weak arrangement of regular and irregular objects. With a higher number of objects, they become loosely stacked in the workspace and start falling or displacing/moving as soon as they are touched by the gripper, leading to failed prehensile and non-prehensile manipulations. Despite these complications, the overall results demonstrate that the agent has achieved a substantial level of learning and convergence to the optimal policy for the pick-and-place task involving regular- and irregular-shaped objects through a sequential combination of prehensile and non-prehensile robotic manipulations.

Discussion

In the experimental trials, we trained and tested our proposed approach agent along with its various variants. A comparative analysis was performed on the results of the proposed approach, its designed variants, and previously presented state-of-the-art methods. The results showing that the proposed approach outperforms the previously published state-of-the-art method by around 5% clearly highlights the positive role of the non-prehensile suction-grip manipulation in the pick-and-place task involving regular and irregular objects in clutter. These results not only demonstrate the agent's success in learning and converging to the optimal policy by understanding the patterns between prehensile and non-prehensile manipulations, but also emphasize the utility of deploying different robotic grippers together. The better performance of the proposed approach compared to the previously presented work underscores the effectiveness of suction gripping and bidirectional sliding manipulations in achieving successful pick-and-place operations. The role of reward-based learning was also explored by designing a binary classification-based agent. The importance of an efficient rewarding scheme and optimal policy convergence through the off-policy model-free Q-learning becomes evident when the binary classification-based variant is restricted to a success rate of under 70%. This variant lacks the reward-based learning ability, which apparently limits its feature learning compared to the proposed approach, resulting in a lower success rate in the pick-and-place task. Another important factor explored was the architecture of the four jointly trained FCNs used for Q-function approximation. It was crucial to determine the role of the FCN architecture in the convergence of the proposed agent to the optimal policy, in order to understand whether the architecture influences Q-function approximation. For this reason, a variant was designed by replacing the DenseNet-121 architecture with the ResNet-101 architecture. Both architectures address the vanishing gradient problem effectively and were chosen for this reason. The role of the FCN architecture in Q-function approximation was confirmed when the proposed approach with DenseNet-121 architecture outperformed the ResNet-101 architecture variant by around 17%. The most likely reason for this outperformance is the difference in feature map handling between the two architectures. DenseNet-121 uses feature concatenation, resulting in higher feature learning and reusability due to its $\boldsymbol{n(n+1)/2}$ connections among layers. In contrast, the ResNet-101 architecture performs feature summation and relies on skip connections. To further investigate the role of non-prehensile manipulations, a variant was designed with zero rewards for the suction-grip, inward-slide, and outward-slide actions. This variant exhibited a slower learning compared to the proposed approach and underperformed in terms of success rate by approximately 26%. This slower learning rate clearly indicates the agent's difficulty in learning synchronization between the prehensile and non-prehensile manipulations. Without rewards encouraging the use of non-prehensile manipulations to maximize future expected rewards, the success rate was lower. This finding supports the earlier results comparing the proposed approach to the previously presented state-of-the-art method. Additionally, the role of depth data was examined by designing a variant that used only RGB channels. As anticipated, performance deteriorated by approximately 41% when only RGB data were provided. This performance decline underscores the importance of depth data in predicting action-values during Q-function approximation. The absence of depth data hindered efficient feature learning by eliminating the concept of distance and spatial relationships. This finding highlights the depth data as a critical element of the proposed approach. As mentioned in previous sections, both pretrained and untrained models of DenseNet-121 architecture were tested.

Surprisingly, the untrained variant performed better in terms of learning pace compared to the pretrained one. This was evident from the earlier rise in the curve of the untrained variant compared to the pretrained one. The likely cause for this could be the difference in pixel patterns between our simulation data and the ImageNet dataset. Thus, the pretrained variant had to first escape the local maxima represented by pretrained weights before it could relearn new weights. Overall, the testing based on clutter density further supports the earlier findings that the agent successfully learned and converged to the optimal policy by sequentially synchronizing non-prehensile and prehensile manipulations, leading to a higher success rate.

6. Conclusions

This research study developed a deep reinforcement learning-based polydexterous framework that enables agents to learn industrial pick-and-place tasks in a self-supervised manner using cobots. The proposed framework allows agents to learn the sequential combination of prehensile and non-prehensile robotic manipulations using different robotic grippers. This learning process facilitates the achievement of optimal policy convergence, leading to increased efficiency and throughput for the pick-and-place task. The framework incorporates four distinct robotic manipulations: grasp, inward-slide, outward-slide, and suction-grip, using two robotic grippers: a jaw gripper and a suction cup. A key contribution of this research is that the framework integrates the manipulations of two different grippers, unlike most existing literature, where a single gripper typically performs all manipulations. Another significant contribution is that the framework operates without requiring prior domain-specific knowledge about objects or the environment, unlike previous approaches. The off-policy model-free Q-learning algorithm is utilized, and the pixelwise-parameterization technique is employed for Q-function approximation. Pixelwise maps of action-values are obtained through pixelwise-parameterization, and the 3D location corresponding to the pixel with the maximum action-value is selected for the relevant robotic manipulation. The reward scheme is designed accordingly, allowing the agent to learn in a completely self-supervised manner by recognizing patterns in non-prehensile manipulations, which increases the likelihood of successful prehensile manipulations in the near future. The Results Section demonstrates that the agents achieve optimal policy convergence, as evidenced by comparisons with baselines, various variants, and density-based testing clutter.

There are some limitations in the presented approach. For instance, the agents are trained on a fixed number of regular- and irregular-shaped objects. This could be extended in future research by incorporating real-life 3D object shapes such as balls, cups, etc. Similarly, transitioning from the simulated environment to the physical world could be achieved by redesigning the approach using the Robot Operating System (ROS/ROS2), which would enable the robotic arm to experiment with real-life objects. Additionally, the sequential combination learning of the agent could be expanded to include parallel combinations of robotic manipulations, incorporating tasks such as rotating and stacking, and exploring their impact on efficiency and throughput.

Author Contributions: M.B.I. conceived and implemented the idea and wrote the paper; Y.Q. and B.L. reviewed and supervised the work. All authors have read and agreed to the published version of the manuscript.

Funding: This publication emanated from research conducted with the support of Science Foundation Ireland (SFI) under Grant Number SFI 16/RC/3918, co-funded by the European Regional Development Fund. Additionally, this work was supported by the North–South Research Programme, which is delivered by the Higher Education Authority (HEA) on behalf of the Department of Further and Higher Education, Research, Innovation, and Science (DFHERIS), and the Shared Island Unit at the Department of the Taoiseach.

Data Availability Statement: Data are available on request from the authors.

Conflicts of Interest: The authors declare no conflicts of interest.

References

1. Vaidya, S.; Ambad, P.; Bhosle, S. Industry 4.0—A Glimpse. *Procedia Manuf.* **2018**, *20*, 233–238. [CrossRef]
2. Friedman, T.L. At Lunch, Donald Trump Gives Critics Hope. *New York Times*, 22 November 2016. Available online: https://www.nytimes.com/2016/11/22/opinion/at-lunch-donald-trump-gives-critics-hope.html (accessed on 1 July 2024).
3. Siciliano, B.; Khatib, O. *Springer Handbook of Robotics*; Springer: Berlin/Heidelberg, Germany, 2007; ISBN 3-540-23957-X.
4. Automation—Computer Process Control | Britannica. Available online: https://www.britannica.com/technology/automation/Robots-in-manufacturing (accessed on 26 January 2022).
5. Bogue, R. Growth in e-commerce boosts innovation in the warehouse robot market. *Ind. Robot Int. J.* **2016**, *43*, 583–587. [CrossRef]
6. Lamiraux, F.; Mirabel, J. Prehensile Manipulation Planning: Modeling, Algorithms and Implementation. *IEEE Trans. Robot.* **2022**, *38*, 2370–2388. [CrossRef]
7. Imtiaz, M.B.; Qiao, Y.; Lee, B. Prehensile Robotic pick-and-place in clutter with Deep Reinforcement Learning. In Proceedings of the 2022 International Conference on Electrical, Computer and Energy Technologies (ICECET), Prague, Czech Republic, 20–22 July 2022; pp. 1–6. [CrossRef]
8. Dogar, M.R.; Srinivasa, S. *A Planning Framework for Non-Prehensile Manipulation under Clutter and Uncertainty*; Carnegie Mellon University: Pittsburgh, PA, USA, 2018. [CrossRef]
9. Imtiaz, M.B.; Qiao, Y.; Lee, B. Prehensile and Non-Prehensile Robotic Pick-and-Place of Objects in Clutter Using Deep Reinforcement Learning. *Sensors* **2023**, *23*, 1513. [CrossRef]
10. Hogan, F.R.; Rodriguez, A. Feedback Control of the Pusher-Slider System: A Story of Hybrid and Underactuated Contact Dynamics. In *Algorithmic Foundations of Robotics XII: Proceedings of the Twelfth Workshop on the Algorithmic Foundations of Robotics*; Goldberg, K., Abbeel, P., Bekris, K., Miller, L., Eds.; Springer International Publishing: Cham, Switzerland, 2020; pp. 800–815, ISBN 978-3-030-43089-4.
11. Clavera, I.; Held, D.; Abbeel, P. Policy transfer via modularity and reward guiding. In Proceedings of the 2017 IEEE/RSJ International Conference on Intelligent Robots and Systems (IROS), Vancouver, BC, Canada, 24–28 September 2017; pp. 1537–1544. [CrossRef]
12. Zeng, A.; Song, S.; Yu, K.-T.; Donlon, E.; Hogan, F.R.; Bauza, M.; Ma, D.; Taylor, O.; Liu, M.; Romo, E.; et al. Robotic pick-and-place of novel objects in clutter with multi-affordance grasping and cross-domain image matching. *Int. J. Robot. Res.* **2022**, *41*, 690–705. [CrossRef]
13. Mahler, J.; Liang, J.; Niyaz, S.; Laskey, M.; Doan, R.; Liu, X.; Ojea, J.A.; Goldberg, K. Dex-Net 2.0: Deep Learning to Plan Robust Grasps with Synthetic Point Clouds and Analytic Grasp Metrics. In Proceedings of the Robotics: Science and Systems XIII, Cambridge, MA, USA, 12–16 July 2017; Amato, N.M., Srinivasa, S.S., Ayanian, N., Kuindersma, S., Eds.; Massachusetts Institute of Technology: Cambridge, MA, USA, 2017. [CrossRef]
14. Koivikko, A.; Drotlef, D.-M.; Sitti, M.; Sariola, V. Magnetically switchable soft suction grippers. *Extreme Mech. Lett.* **2021**, *44*, 101263. [CrossRef]
15. Rodriguez, A.; Mason, M.T.; Ferry, S. From caging to grasping. *Int. J. Robot. Res.* **2012**, *31*, 886–900. [CrossRef]
16. Singh, T.; Ambike, S. A soft-contact and wrench based approach to study grasp planning and execution. *J. Biomech.* **2015**, *48*, 3961–3967. [CrossRef]
17. Zhou, J.; Liu, Y.; Liu, J.; Xie, Q.; Zhang, Y.; Zhu, X.; Ding, X. BOLD3D: A 3D BOLD descriptor for 6Dof pose estimation. *Comput. Graph.* **2020**, *89*, 94–104. [CrossRef]
18. Goldfeder, C.; Ciocarlie, M.; Dang, H.; Allen, P.K. The Columbia grasp database. In Proceedings of the 2009 IEEE International Conference on Robotics and Automation, Kobe, Japan, 12–17 May 2009; pp. 1710–1716.
19. Pinto, L.; Gupta, A. Learning to push by grasping: Using multiple tasks for effective learning. In Proceedings of the 2017 IEEE International Conference on Robotics and Automation (ICRA), Singapore, 29 May–3 June 2017; pp. 2161–2168. [CrossRef]
20. Lynch, K.M. Estimating the friction parameters of pushed objects. In Proceedings of the 1993 IEEE/RSJ International Conference on Intelligent Robots and Systems (IROS '93), Yokohama, Japan, 26–30 July 1993; Volume 1, pp. 186–193. [CrossRef]
21. Bauza, M.; Rodriguez, A. A probabilistic data-driven model for planar pushing. In Proceedings of the 2017 IEEE International Conference on Robotics and Automation (ICRA), Singapore, 29 May–3 June 2017; pp. 3008–3015. [CrossRef]
22. Zhou, J.; Mason, M.T.; Paolini, R.; Bagnell, D. A convex polynomial model for planar sliding mechanics: Theory, application, and experimental validation. *Int. J. Robot. Res.* **2018**, *37*, 249–265. [CrossRef]
23. Owen-Hill, A. Why Vacuum Grippers Are Really the Best Option for Robot Palletizing. Available online: https://blog.robotiq.com/why-vacuum-grippers-are-really-the-best-option-for-robot-palletizing (accessed on 2 July 2024).
24. How Does a Vacuum Gripper Work? Available online: https://www.unboxindustry.com/blog/7-how-does-a-vacuum-gripper-work (accessed on 3 July 2024).
25. Eppner, C.; Höfer, S.; Jonschkowski, R.; Martín-Martín, R.; Sieverling, A.; Wall, V.; Brock, O. Lessons from the Amazon picking challenge: Four aspects of building robotic systems. In Proceedings of the 26th International Joint Conference on Artificial Intelligence, IJCAI'17, Melbourne, Australia, 19–25 August 2017; AAAI Press: Melbourne, Australia, 2017; pp. 4831–4835.
26. Hernandez, C.; Bharatheesha, M.; Ko, W.; Gaiser, H.; Tan, J.; van Deurzen, K.; de Vries, M.; Van Mil, B.; van Egmond, J.; Burger, R.; et al. Team Delft's Robot Winner of the Amazon Picking Challenge 2016. In Proceedings of the RoboCup 2016: Robot World Cup XX; Behnke, S., Sheh, R., Sarıel, S., Lee, D.D., Eds.; Springer International Publishing: Cham, Switzerland, 2017; pp. 613–624.

27. Schwarz, M.; Milan, A.; Lenz, C.; Muñoz, A.; Periyasamy, A.S.; Schreiber, M.; Schüller, S.; Behnke, S. NimbRo picking: Versatile part handling for warehouse automation. In Proceedings of the 2017 IEEE International Conference on Robotics and Automation (ICRA), Singapore, 29 May–3 June 2017; pp. 3032–3039. [CrossRef]
28. Domae, Y.; Okuda, H.; Taguchi, Y.; Sumi, K.; Hirai, T. Fast graspability evaluation on single depth maps for bin picking with general grippers. In Proceedings of the 2014 IEEE International Conference on Robotics and Automation (ICRA), Hong Kong, China, 31 May–7 June 2014; pp. 1997–2004. [CrossRef]
29. Mahler, J.; Matl, M.; Liu, X.; Li, A.; Gealy, D.; Goldberg, K. Dex-Net 3.0: Computing Robust Vacuum Suction Grasp Targets in Point Clouds using a New Analytic Model and Deep Learning. In Proceedings of the 2018 IEEE International Conference on Robotics and Automation (ICRA), Brisbane, QLD, Australia, 21–25 May 2018.
30. Liu, W.; Pan, Z.; Liu, W.; Shao, Q.; Hu, J.; Wang, W.; Ma, J.; Qi, J.; Zhao, W.; Du, S. Deep learning for picking point detection in dense cluster. In Proceedings of the 2017 11th Asian Control Conference (ASCC), Gold Coast, QLD, Australia, 17–20 December 2017; pp. 1644–1649. [CrossRef]
31. Boularias, A.; Bagnell, J.; Stentz, A. Learning to Manipulate Unknown Objects in Clutter by Reinforcement. In Proceedings of the Twenty-Ninth AAAI Conference on Artificial Intelligence, Austin, TX, USA, 25–30 January 2015; Volume 29. [CrossRef]
32. Sarantopoulos, I.; Kiatos, M.; Doulgeri, Z.; Malassiotis, S. Total Singulation with Modular Reinforcement Learning. *IEEE Robot. Autom. Lett.* **2021**, *6*, 4117–4124. [CrossRef]
33. Imtiaz, M.B.; Qiao, Y.; Lee, B. Implementing Robotic Pick and Place with Non-visual Sensing Using Reinforcement Learning. In Proceedings of the 2022 6th International Conference on Robotics, Control and Automation (ICRCA), Xiamen, China, 26–28 February 2022; pp. 23–28. [CrossRef]
34. Imtiaz, M.B.; Qiao, Y.; Lee, B. Comparison of Two Reinforcement Learning Algorithms for Robotic Pick and Place with Non-Visual Sensing. *Int. J. Mech. Eng. Robot. Res.* **2021**, *10*, 526–535. [CrossRef]
35. Pitts, W.; McCulloch, W.S. How we know universals the perception of auditory and visual forms. *Bull. Math. Biophys.* **1947**, *9*, 127–147. [CrossRef]
36. Krizhevsky, A.; Sutskever, I.; Hinton, G.E. ImageNet Classification with Deep Convolutional Neural Networks. In Proceedings of the 25th International Conference on Neural Information Processing Systems—Volume 1, NIPS'12, Lake Tahoe, NV, USA, 3–6 December 2012; Curran Associates Inc.: Red Hook, NY, USA, 2012; pp. 1097–1105.
37. Mohammed, M.Q.; Kwek, L.C.; Chua, S.C.; Al-Dhaqm, A.; Nahavandi, S.; Eisa, T.A.E.; Miskon, M.F.; Al-Mhiqani, M.N.; Ali, A.; Abaker, M.; et al. Review of learning-based robotic manipulation in cluttered environments. *Sensors* **2022**, *22*, 7938. [CrossRef]
38. Kaelbling, L.P.; Littman, M.L.; Moore, A.W. Reinforcement Learning: A Survey. *J. Artif. Int. Res.* **1996**, *4*, 237–285. [CrossRef]
39. Kober, J.; Bagnell, J.A.; Peters, J. Reinforcement learning in robotics: A survey. *Int. J. Robot. Res.* **2013**, *32*, 1238–1274. [CrossRef]
40. Sutton, R.S.; Barto, A.G. *Reinforcement Learning: An Introduction*; MIT Press: Cambridge, MA, USA, 2018; p. 352.
41. Hessel, M.; Modayil, J.; van Hasselt, H.; Schaul, T.; Ostrovski, G.; Dabney, W.; Horgan, D.; Piot, B.; Azar, M.G.; Silver, D. Rainbow: Combining Improvements in Deep Reinforcement Learning. *arXiv* **2017**, arXiv:1710.02298. [CrossRef]
42. Mnih, V.; Kavukcuoglu, K.; Silver, D.; Graves, A.; Antonoglou, I.; Wierstra, D.; Riedmiller, M.A. Playing Atari with Deep Reinforcement Learning. *arXiv* **2013**, arXiv:1312.5602.
43. Mnih, V.; Kavukcuoglu, K.; Silver, D.; Rusu, A.A.; Veness, J.; Bellemare, M.G.; Graves, A.; Riedmiller, M.; Fidjeland, A.K.; Ostrovski, G.; et al. Human-level control through deep reinforcement learning. *Nature* **2015**, *518*, 529–533. [CrossRef]
44. Long, J.; Shelhamer, E.; Darrell, T. Fully convolutional networks for semantic segmentation. In Proceedings of the 2015 IEEE Conference on Computer Vision and Pattern Recognition (CVPR), Boston, MA, USA, 7–12 June 2015; IEEE Computer Society: Los Alamitos, CA, USA, 2015; pp. 3431–3440. [CrossRef]
45. Huang, G.; Liu, Z.; van der Maaten, L.; Weinberger, K.Q. Densely Connected Convolutional Networks. *arXiv* **2018**, arXiv:1608.06993.
46. Huber, P.J. Robust Estimation of a Location Parameter. *Ann. Math. Stat.* **1964**, *35*, 492–518. [CrossRef]
47. Liu, Y.; Gao, Y.; Yin, W. An Improved Analysis of Stochastic Gradient Descent with Momentum. *arXiv* **2020**, arXiv:2007.07989.
48. Pleiss, G.; Chen, D.; Huang, G.; Li, T.; van der Maaten, L.; Weinberger, K.Q. Memory-Efficient Implementation of DenseNets. *arXiv* **2017**, arXiv:1707.06990.
49. Diankov, R.; Kuffner, J.J. *OpenRAVE: A Planning Architecture for Autonomous Robotics*; Robotics Institute: Pittsburgh, PA, USA, 2008.
50. Schulman, J.; Ho, J.; Lee, A.X.; Awwal, I.; Bradlow, H.; Abbeel, P. Finding Locally Optimal, Collision-Free Trajectories with Sequential Convex Optimization. In Proceedings of the Robotics: Science and Systems IX, Berlin, Germany, 24–28 June 2013.
51. Şucan, I.A.; Moll, M.; Kavraki, L.E. The Open Motion Planning Library. *IEEE Robot. Autom. Mag.* **2012**, *19*, 72–82. [CrossRef]
52. Kuffner, J.J.; LaValle, S.M. RRT-connect: An efficient approach to single-query path planning. In *Proceedings of the 2000 ICRA. Millennium Conference. In Proceedings of the IEEE International Conference on Robotics and Automation. Symposia Proceedings (Cat. No. 00CH37065)*; San Francisco, CA, USA, 24–28 April 2000, Volume 2, pp. 995–1001. [CrossRef]
53. He, K.; Zhang, X.; Ren, S.; Sun, J. Deep Residual Learning for Image Recognition. *arXiv* **2020**, arXiv:1512.03385.

MDPI AG
Grosspeteranlage 5
4052 Basel
Switzerland
Tel.: +41 61 683 77 34

Machines Editorial Office
E-mail: machines@mdpi.com
www.mdpi.com/journal/machines